LECKIE
the education publisher
for Scotland

National 4 to 5
MATHS

Bridging Skills Book
Craig Lowther, Clare Ford
and Dominic Kennedy

© 2018 Leckie
cover image © TukkataMoji / Shutterstock

001/09082018

10 9 8 7 6 5 4 3 2

ISBN 9780008209063

Published by
Leckie
An imprint of HarperCollins*Publishers*
Westerhill Road, Bishopbriggs, Glasgow, G64 2QT
T: 0844 576 8126   F: 0844 576 8131
leckiescotland@harpercollins.co.uk
www.leckiescotland.co.uk

HarperCollins Publishers
Macken House, 39/40 Mayor Street Upper, Dublin 1, D01 C9W8, Ireland

Special thanks to
Jouve (layout and illustration)
Ink Tank (cover design)
Project One Publishing Solutions (project management and editing)
Rachel Hamar (proof read)
Nick Hamar (answer check)

A CIP Catalogue record for this book is available from the British Library.

Acknowledgements

Whilst every effort has been made to trace the copyright holders, in
cases where this has been unsuccessful, or if any have inadvertently been
overlooked, the Publishers would gladly receive any information enabling
them to rectify any error or omission at the first opportunity.

Printed in the UK.

This book contains FSC™ certified paper and other controlled
sources to ensure responsible forest management.

For more information visit: www.harpercollins.co.uk/green

Contents

Answers are provided online at:
https://collins.co.uk/pages/scottish-curriculum-free-resources

N4 **Example 1.2**

Evaluate:

a $0 + (-8)$ b $-10 - (-2)$ c $12 + (-13) - 2 - (-5)$

a $0 + (-8) = 0 - 8$ •——————— | When you add a negative number, you subtract. |

$\quad\quad\quad = -8$

b $-10 - (-2) = -10 + 2$ •——————— | When you subtract a negative number, you add. |

$\quad\quad\quad = -8$

c $12 + (-13) - 2 - (-5) = 12 - 13 - 2 + 5$

$\quad\quad\quad = 2$

| Hint | Sometimes a negative number is written with brackets around it, for example, (-8). You don't have to do this, but it makes it easier to keep track of what you are doing. |

N4 **Exercise 1A**

1 Calculate:

    a $2 - 9$ b $21 - 35$ c $-3 - 10$ d $-250 - 150$

    e $-6 + 6$ f $-80 + 40$ g $-15 + 25$ h $-12 + 18$

★ 2 Find:

    a $8 + (-2)$ b $53 + (-34)$ c $12 + (-12)$

    d $21 + (-30)$ e $56 + (-100)$ f $-6 + (-3)$

    g $-42 + (-50)$ h $4 - (-3)$ i $29 - (-31)$

    j $-16 - (-20)$ k $-100 - (-100)$ l $-56 - (-23)$

    m $10 + (-8) - (-5)$ n $-16 - (-2) + (-40)$ o $-1 + (-2) - (-3) - (-10) + 6$

Multiplication and division revision

The table below provides a reminder of the patterns for multiplying or dividing integers.

First number	Operation	Second number	Answer
positive	× or ÷	positive	positive
positive	× or ÷	negative	negative
negative	× or ÷	positive	negative
negative	× or ÷	negative	positive

N4 **Example 1.3**

Find :

a $25 \times (-4)$ b $\dfrac{-3}{-6}$ c $(-4)^2$ d $\dfrac{200}{5 \times (-2)}$

a $25 \times (-4) = -100$ •——————— | Positive × negative → negative |

b $\dfrac{-3}{-6} = \dfrac{1}{2}$ •——————— | Negative ÷ negative → positive |

c $(-4)^2 = (-4) \times (-4)$ ────────── To square a number means to multiply it by itself.

   $= 16$ ────────── Negative × negative → positive

d $\dfrac{200}{5 \times (-2)} = \dfrac{200}{-10}$ ────────── Multiply the numbers in the denominator first.

   $= -20$ ────────── Positive ÷ negative → negative

N4 ## Exercise 1B

★ 1 Find:

a 12×10 b 42×8 c $7 \div 7$ d $\dfrac{48}{12}$

e $15 \times (-2)$ f -100×3 g $-20 \div 4$ h $\dfrac{45}{-5}$

i $-4 \times (-2)$ j $(-12) \times (-5)$ k $\dfrac{-30}{-5}$ l $(-2) \div (-2)$

> **Hint** Look at the table on page 2 if you are unsure whether your final answer is positive or negative.

2 Calculate:

a 21×0 b $0 \times (-5)$ c $(-5) \div 10$ d $(-25) \times (-8)$

e $40 \div (-4)$ f $\dfrac{0}{-4}$ g 1^2 h $(-5)^2$

i $\dfrac{8 \times (-3)}{6}$ j $\dfrac{16}{2 \times (-2)}$ k $\dfrac{-32}{8 \times (-1)}$ l $\dfrac{(-2) \times (-9)}{(-3)^2}$

N5 # Numerical calculations with fractions

When carrying out numerical calculations – addition, subtraction, multiplication and division – with fractions, you need to know the following terms:

- the number (or letter or variable) on the top of a fraction is called the **numerator**

- the number (or letter or variable) on the bottom of the fraction is called the **denominator**

  $\dfrac{2}{3}$ ←── numerator
     ←── denominator

- fractions such as $\dfrac{1}{2}$ and $\dfrac{3}{4}$, where the numerator is less than the denominator, are called **proper fractions**

- fractions such as $\dfrac{3}{2}$ and $\dfrac{17}{5}$, where the numerator is greater than the denominator, are called **improper fractions**

- numbers such as $4\frac{5}{6}$, with a whole number part and a fraction part, are called **mixed numbers**.

Adding and subtracting fractions

In order to add or subtract fractions, the fractions must have a **common denominator** (that is, the numbers on the bottom of the fractions must be the same).

When you add two fractions with a common denominator, you add the numerators, for example:

$$\frac{1}{9} + \frac{4}{9} = \frac{1+4}{9} = \frac{5}{9}$$

Similarly, when you subtract two fractions with a common denominator, you subtract the second numerator from the first, for example:

$$\frac{7}{8} - \frac{3}{8} = \frac{7-3}{8} = \frac{4}{8} = \frac{1}{2}$$

In this example, $\frac{4}{8}$ can be simplified to $\frac{1}{2}$

If the fractions do not have a common denominator, you need to convert one or both of them so they do have a common denominator. To identify a common denominator you can multiply the denominators. In order that you don't change the value of the fraction you must multiply the numerator by the same number as you multiply the denominator. For example,

$$\frac{2}{3} + \frac{1}{5} = \frac{2 \times 5}{3 \times 5} + \frac{1 \times 3}{5 \times 3}$$

$$= \frac{10}{15} + \frac{3}{15} = \frac{13}{15}$$

If the fractions being added or subtracted do not have a common denominator but one is a multiple of the other, you only need to convert one of them. For example, to work out $\frac{3}{8} - \frac{1}{4}$, you could multiply the denominators (4 and 8) to obtain a common denominator (32), but this is not the **lowest** common denominator. It is more straightforward to simply convert $\frac{1}{4}$ so it has a denominator of 8:

$$\frac{3}{8} - \frac{1}{4} = \frac{3}{8} - \frac{1 \times 2}{4 \times 2}$$

$$= \frac{3}{8} - \frac{2}{8} = \frac{1}{8}$$

Converting improper fractions to mixed numbers

To convert an improper fraction into a mixed number, split the improper fraction to make a fraction that will simplify into a whole number and a second proper fraction, then combine them. For example:

$$\frac{11}{10} = \frac{10}{10} + \frac{1}{10}$$

$$= 1 + \frac{1}{10} = 1\frac{1}{10}$$

Similarly:

$$\frac{8}{3} = \frac{6}{3} + \frac{2}{3}$$

$$= 2 + \frac{2}{3} = 2\frac{2}{3}$$

N5

Example 1.4

Find:

a $\dfrac{2}{5} + \dfrac{3}{4}$
b $\dfrac{5}{6} - \dfrac{1}{4}$
c $1\dfrac{2}{3} + 4\dfrac{3}{7}$

a $\dfrac{2}{5} + \dfrac{3}{4} = \dfrac{2 \times 4}{5 \times 4} + \dfrac{3 \times 5}{4 \times 5}$

First, convert both fractions so they have a common denominator. 20 is the lowest common multiple of 4 and 5.

$= \dfrac{8}{20} + \dfrac{15}{20} = \dfrac{8 + 15}{20}$

Add the numerators.

$= \dfrac{23}{20} = \dfrac{20}{20} + \dfrac{3}{20}$

$= 1\dfrac{3}{20}$

Convert the improper fraction to a mixed number.

b $\dfrac{5}{6} - \dfrac{1}{4} = \dfrac{5 \times 2}{6 \times 2} - \dfrac{1 \times 3}{4 \times 3}$

12 is the lowest common multiple of 6 and 4.

$= \dfrac{10}{12} - \dfrac{3}{12} = \dfrac{10 - 3}{12}$

Subtract the numerators.

$= \dfrac{7}{12}$

c $1\dfrac{2}{3} + 4\dfrac{3}{7} = 5 + \dfrac{2}{3} + \dfrac{3}{7}$

Add the whole numbers and the fraction parts separately.

$= 5 + \dfrac{2 \times 7}{3 \times 7} + \dfrac{3 \times 3}{7 \times 3}$

21 is the lowest common multiple of 3 and 7.

$= 5 + \dfrac{14}{21} + \dfrac{9}{21}$

$= 5 + \dfrac{14 + 9}{21}$

Add the numerators.

$= 5 + \dfrac{23}{21} = 5 + \dfrac{21}{21} + \dfrac{2}{21}$

Begin to convert the improper fraction to a mixed number.

$= 5 + 1 + \dfrac{2}{21} = 6\dfrac{2}{21}$

Add the whole numbers.

N5

Exercise 1C

1 Find:

a $\dfrac{2}{5} + \dfrac{1}{5}$
b $\dfrac{10}{11} - \dfrac{3}{11}$
c $\dfrac{7}{8} - \dfrac{5}{8}$
d $\dfrac{8}{9} + \dfrac{5}{9}$

e $\dfrac{5}{6} - \dfrac{1}{6}$
f $6\dfrac{4}{5} - 2\dfrac{1}{5}$
g $8\dfrac{2}{3} - \dfrac{1}{3}$
h $5\dfrac{7}{9} - 5\dfrac{1}{9}$

i $1\dfrac{1}{4} + 3\dfrac{1}{4}$
j $4\dfrac{2}{5} + 2\dfrac{3}{5}$
k $3\dfrac{8}{9} + \dfrac{2}{9}$
l $4\dfrac{3}{7} - \dfrac{4}{7}$

2 Calculate:

a $\frac{2}{3} + \frac{1}{5}$ b $\frac{3}{4} - \frac{1}{3}$ c $\frac{2}{7} + \frac{2}{5}$ d $\frac{8}{9} - \frac{3}{4}$

e $\frac{4}{5} + \frac{2}{3}$ f $\frac{5}{9} + \frac{3}{4}$ g $\frac{5}{6} + \frac{2}{3}$ h $\frac{7}{8} - \frac{1}{2}$

i $\frac{4}{5} - \frac{3}{10}$ j $\frac{5}{6} + \frac{1}{8}$ k $\frac{14}{15} - \frac{5}{6}$ l $\frac{1}{12} + \frac{1}{16}$

★ 3 Work out:

a $1\frac{3}{5} + 2\frac{1}{2}$ b $3\frac{11}{12} - 3\frac{4}{5}$ c $5\frac{1}{2} - 4\frac{1}{5}$ d $4\frac{3}{4} - 1\frac{2}{5}$

e $4\frac{5}{6} + \frac{2}{3}$ f $6\frac{3}{8} - \frac{1}{12}$ g $3\frac{4}{7} + 3\frac{3}{14}$ h $10\frac{1}{10} + 2\frac{1}{2}$

Multiplying and dividing fractions

Multiplying fractions

These diagrams illustrate that $\frac{1}{2}$ of $\frac{1}{2}$ is $\frac{1}{4}$:

one half $\left(\frac{1}{2}\right)$ one half of one half is one quarter $\left(\frac{1}{2}\text{ of }\frac{1}{2} = \frac{1}{4}\right)$

 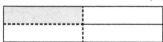

Similarly, these diagrams show that $\frac{1}{3}$ of $\frac{1}{2}$ is $\frac{1}{6}$:

one half $\left(\frac{1}{2}\right)$ one third of one half is one sixth $\left(\frac{1}{3}\text{ of }\frac{1}{2} = \frac{1}{6}\right)$

 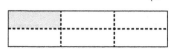

In maths, a fraction 'of' an amount means to multiply the amount by the fraction, so:

$$\frac{1}{2}\text{ of }\frac{1}{2} = \frac{1}{2} \times \frac{1}{2} \qquad\qquad \text{and} \qquad\qquad \frac{1}{3}\text{ of }\frac{1}{2} = \frac{1}{3} \times \frac{1}{2}$$

$$= \frac{1}{4} \qquad\qquad\qquad\qquad\qquad\qquad\qquad = \frac{1}{6}$$

This illustrates the following rule for multiplying fractions:

> **Important**
>
> To multiply two fractions together, multiply the numerators together and multiply the denominators together.

So, for example:

$$\frac{2}{3} \times \frac{4}{5} = \frac{2 \times 4}{3 \times 5} = \frac{8}{15}$$

Dividing by a fraction

These diagrams illustrate that $6 \div 2 = 3$:

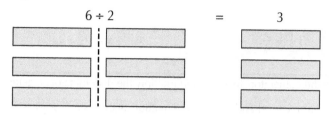

But what is $6 \div \frac{1}{2}$? Using diagrams again:

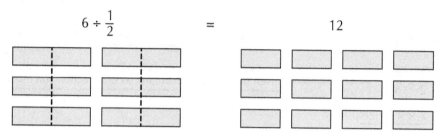

Similarly, these diagrams show that $2 \div \frac{1}{4} = 8$:

$$6 \div \frac{1}{2} = 12 \qquad \text{and} \qquad 2 \div \frac{1}{4} = 8$$

These examples illustrate the following rule for dividing by a fraction:

> **Important**
>
> To divide by a fraction, invert the fraction (turn it upside down) and multiply.

Applying the rule to the examples above:

$$6 \div \frac{1}{2} = \frac{6}{1} \div \frac{1}{2} \qquad \text{and} \qquad 2 \div \frac{1}{4} = \frac{2}{1} \div \frac{1}{4}$$

$$= \frac{6}{1} \times \frac{2}{1} = \frac{6 \times 2}{1 \times 1} \qquad\qquad = \frac{2}{1} \times \frac{4}{1} = \frac{2 \times 4}{1 \times 1}$$

$$= \frac{12}{1} = 12 \qquad\qquad\qquad = \frac{8}{1} = 8$$

And applying the rule to a new example:

$$\frac{3}{5} \div \frac{2}{7} = \frac{3}{5} \times \frac{7}{2} = \frac{3 \times 7}{5 \times 2} = \frac{21}{10}$$

$$= \frac{20}{10} + \frac{1}{10} = 2 + \frac{1}{10} = 2\frac{1}{10}$$

Notice that a common denominator is not needed when you multiply and divide fractions (unlike when adding and subtracting).

N5 **Example 1.5**

Find:

a $\frac{2}{3} \times \frac{5}{6}$

b $3\frac{1}{2} \div \frac{3}{4}$

c $4\frac{2}{5} \div 1\frac{1}{3}$

a $\frac{2}{3} \times \frac{5}{6} = \frac{2 \times 5}{3 \times 6}$

Multiply the numerators together and multiply the denominators together.
You could also cancel and simplify the multiplication first: $\frac{\cancel{2}^{1}}{3} \times \frac{5}{\cancel{6}_{3}} = \frac{1}{3} \times \frac{5}{3}$

$= \frac{10}{18} = \frac{5}{9}$

Simplify the fraction.

b $3\frac{1}{2} \div \frac{3}{4} = \frac{7}{2} \div \frac{3}{4}$

Convert the mixed number to an improper fraction.

$= \frac{7}{2} \times \frac{4}{3}$

Invert (turn upside down) the second fraction and multiply.

$= \frac{7 \times 4}{2 \times 3} = \frac{28}{6}$

$= \frac{14}{3} = \frac{12}{3} + \frac{2}{3}$

Begin to convert the improper fraction to a mixed number.

$= 4\frac{2}{3}$

c $4\frac{2}{5} \div 1\frac{1}{3} = \frac{22}{5} \div \frac{4}{3}$

Convert both mixed numbers to improper fractions.

$= \frac{22}{5} \times \frac{3}{4} = \frac{22 \times 3}{5 \times 4} = \frac{66}{20}$

$= \frac{33}{10} = \frac{30}{10} + \frac{3}{10}$

$= 3\frac{3}{10}$

N5 **Exercise 1D**

1 Find:

a $\frac{3}{5} \times \frac{1}{2}$ b $\frac{3}{7} \times \frac{3}{4}$ c $\frac{2}{5} \div \frac{3}{7}$ d $\frac{3}{11} \div \frac{1}{2}$

e $\frac{6}{7} \times \frac{2}{3}$ f $\frac{4}{9} \times \frac{3}{8}$ g $\frac{5}{8} \div \frac{3}{4}$ h $\frac{4}{9} \div \frac{2}{3}$

★ 2 Work out:

a $1\frac{2}{3} \times \frac{1}{2}$ b $\frac{4}{5} \times 4\frac{1}{3}$ c $2\frac{2}{3} \div \frac{4}{5}$ d $\frac{7}{8} \div 1\frac{3}{4}$

3 Calculate:

a $2\frac{1}{2} \times 3\frac{2}{3}$ b $1\frac{3}{4} \times 1\frac{2}{5}$ c $2\frac{2}{3} \div 3\frac{1}{2}$ d $6\frac{1}{2} \div 2\frac{3}{4}$

N4 ▶N5 ## Numerical calculations with indices

Powers of numbers

5^2 is a shorthand way of writing 5×5. In the expression 5^2, 5 is the **base** and 2 is the **power** or **index**. The plural of index is **indices**. So, when working with indices, you are working with numbers or expressions which contain powers.

$$\text{base} \longrightarrow 5^2 \longleftarrow \text{index}$$

The index shows how many times the base is multiplied by itself.

So:

5^2 (read as '5 squared') $= 5 \times 5 = 25$

2^3 (read as '2 cubed') $= 2 \times 2 \times 2 = 8$

3^5 (read as '3 to the power 5') $= 3 \times 3 \times 3 \times 3 \times 3 = 243$

N4 ▶N5 ## Example 1.6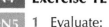

Evaluate:

a 9^2 b 5^3 c 1^6 d $(-10)^2$ e $(-2)^3$ f $4^3+(-1)^2-3^3$

a $9^2 = 9 \times 9 = 81$

b $5^3 = 5 \times 5 \times 5 = 125$

c $1^6 = 1 \times 1 \times 1 \times 1 \times 1 \times 1 = 1$ ●───────── $1 \times 1 = 1$

d $(-10)^2 = (-10) \times (-10) = 100$ ●───── Negative × negative → positive

e $(-2)^3 = (-2) \times (-2) \times (-2)$

   $= 4 \times (-2) = -8$ ●───── Or you could do this in a single step, working out $2 \times 2 \times 2 = 8$, and then applying the rules of multiplying negative numbers: negative × negative → positive, then positive × negative → negative

f $4^3 + (-1)^2 - 3^3 = 4 \times 4 \times 4 + (-1) \times (-1) - 3 \times 3 \times 3$

   $= 64 + 1 - 27 = 38$

N4 ▶N5 ## Exercise 1E

1 Evaluate:

   a 6^2 b 2^2 c 1^2 d 0^2 e 8^2 f 30^2

2 Find:

   a 3^3 b 10^3 c 2^4 d 1^5 e 0^6 f 3^4

3 Calculate:

   a $(-5)^2$ b $(-7)^2$ c $(-3)^3$ d $(-4)^3$

★ 4 Work out:

   a 3^2+2^3 b 1^4+10^3 c 5^3-5^2 d 4^2-2^4

   e $(-3)^2+3^3$ f $10^3-(-1)^2$ g $9^2+(-2)^3-5$ h $(-10)^2+(-1)^3+(-6)$

Roots of numbers

The **square root** of a number is the number which, when multiplied by itself, gives the original number. For example, the square root of 9 is 3 because $3 \times 3 = 9$.

The symbol $\sqrt{\ }$ is used for square root. So, $\sqrt{9} = 3$.

The square root of 9 is also -3 because $-3 \times -3 = 9$. However, for the purposes of this section and calculations with surds (see Chapter 5), we will take $\sqrt{9}$ to mean the positive (principal) square root of 9. So, $\sqrt{9} = 3$, $\sqrt{4} = 2$, etc.

The **cube root** of a number is the number which, when multiplied by itself three times, gives the original number. For example, the cube root of 8 is 2, since $2 \times 2 \times 2 = 8$.

The symbol $\sqrt[3]{\ }$ is used for cube root. So, $\sqrt[3]{8} = 2$.

Similarly, $\sqrt[3]{1000} = 10$ (because $10 \times 10 \times 10 = 1000$).

The **fourth root** $\left(\sqrt[4]{\ }\right)$ of a number is the number which, when multiplied by itself 4 times, gives the original number. For example, the fourth root of 81 is 3, because $3 \times 3 \times 3 \times 3 = 81$.
The symbol $\sqrt[4]{\ }$ is used for the fourth root. So, $\sqrt[4]{81} = 3$.
Similarly, $\sqrt[4]{1} = 1$, because $1 \times 1 \times 1 \times 1 = 1$.

N4
▶N5

Example 1.7

Evaluate these roots:

a $\sqrt{49}$ b $\sqrt{900}$ c $\sqrt[3]{125}$ d $\sqrt[4]{16}$ e $\sqrt[6]{1\,000\,000}$

a $\sqrt{49} = 7$ •————($7 \times 7 = 49$)

b $\sqrt{900} = 30$ •————($30 \times 30 = 900$)

c $\sqrt[3]{125} = 5$ •————($5 \times 5 \times 5 = 125$)

d $\sqrt[4]{16} = 2$ •————($2 \times 2 \times 2 \times 2 = 16$)

e $\sqrt[6]{1\,000\,000} = 10$ •————($10 \times 10 \times 10 \times 10 \times 10 \times 10 = 1\,000\,000$)

N4
▶N5

Exercise 1F

★ 1 Evaluate:

a $\sqrt{36}$ b $\sqrt{4}$ c $\sqrt{1}$ d $\sqrt{1600}$

e $\sqrt{144}$ f $\sqrt{169}$ g $\sqrt{225}$ h $\sqrt{2500}$

★ 2 Find:

a $\sqrt[3]{27}$ b $\sqrt[3]{1000}$ c $\sqrt[3]{1}$ d $\sqrt[3]{-8000}$

e $\sqrt[3]{64}$ f $\sqrt[3]{0}$ g $\sqrt[3]{27\,000}$ h $\sqrt[3]{-1}$

★ 3 Calculate the following.

a $\sqrt[4]{10\,000}$ b $\sqrt[5]{1}$ c $\sqrt[4]{625}$ d $\sqrt[5]{32}$

e $\sqrt[5]{100\,000}$ f $\sqrt[3]{8}$ g $\sqrt[7]{1}$ h $\sqrt[10]{1}$

N4 **Chapter 1 review** 🔀

▶N5

1 Find:

a $8 - 12$ b $-3 - 10$ c $-7 + 8$ d $-12 + 1$

e $10 + (-2)$ f $8 + (-15)$ g $-4 + (-6)$ h $50 - (-3)$

i $-7 - (-6)$ j $20 + (-5) - (-3)$ k $-100 - (-200) + (-300)$

2 Work out:

a 23×15 b $\dfrac{30}{30}$ c $6 \times (-4)$ d $(-3) \times 10$

e $20 \div (-4)$ f $\dfrac{-100}{5}$ g $(-4) \times (-3)$ h $(-5)^2$

i $\dfrac{-8}{-4}$ j $(-6) \div (-6)$ k $\dfrac{5 \times (-6)}{(-3)}$

3 Calculate:

a $\dfrac{2}{3} + \dfrac{1}{2}$ b $\dfrac{7}{8} - \dfrac{3}{5}$ c $4\dfrac{1}{5} + \dfrac{3}{4}$ d $5\dfrac{7}{10} - 2\dfrac{3}{7}$

4 Evaluate:

a $\dfrac{2}{3} \times \dfrac{3}{5}$ b $\dfrac{3}{7} \div \dfrac{4}{5}$ c $2\dfrac{1}{5} \times \dfrac{3}{8}$ d $3\dfrac{1}{2} \div 1\dfrac{1}{3}$

5 Work out:

a $(-3)^2$ b 2^3 c 1^4 d 0^5

e 10^6 f $4^3 - 4^2$ g $(-5)^2 + 5^3$ h $(-9)^2 - (-1)^3 + (-4)$

6 Evaluate the following roots.

a $\sqrt{64}$ b $\sqrt[3]{8}$ c $\sqrt{289}$

d $\sqrt[4]{625}$ e $\sqrt[6]{1}$ f $\sqrt[7]{10\,000\,000}$

- I can carry out basic calculations involving integers
  (positive numbers, negative numbers and zero).
  ★ Exercise 1A Q2 ★ Exercise 1B Q1

- I can carry out numerical calculations involving fractions
  (addition, subtraction, multiplication and division).
  ★ Exercise 1C Q3 ★ Exercise 1D Q2

- I can carry out numerical calculations with basic
  indices (or powers) and roots. ★ Exercise 1E Q4
  ★ Exercise 1F Q1–Q3

2 Algebra 1

This chapter will show you how to:

- work with algebraic expressions involving expansion of brackets
- factorise an algebraic expression (involving a common factor, the difference of two squares or a trinomial)
- reduce an algebraic fraction to its simplest form
- apply one of the four operations $(+, -, \times, \div)$ to algebraic fractions.

You should already know how to:

- expand a pair of brackets by multiplying through by a number, for example, $4(3x - 1) = 12x - 4$
- factorise an expression by extracting a numerical common factor, for example, $16a + 24 = 8(2a + 3)$
- simplify an expression which has more than one variable (or letter) by collecting like terms, for example, $6x + 3y - 5x + y = x + 4y$
- evaluate an expression or a formula which has more than one variable, for example, when $s = 1$ and $t = 3$, $5s + 2t = 5 \times 1 + 2 \times 3 = 11$

N4 Algebra revision

This section revises the following algebra topics:

- expanding a pair of brackets with a number at the front
- factorising an expression by taking out a numerical common factor
- simplifying expressions with one or more variables (or letters)
- finding the value of an expression or a formula given specific values.

To **expand** a pair of brackets, **everything inside** the bracket is multiplied separately by whatever is in front of the bracket. So $2(3x - 1) = 2 \times 3x + 2 \times (-1)$.

> **Hint** When using brackets, the sign for multiplication, $\times$, is usually omitted, for example, $2 \times (3x - 1)$ is normally written as $2(3x - 1)$.
>
> The terms inside the brackets keep the sign in front of them.

To **factorise** is the opposite of to expand.

To **simplify** means to 'collect like terms' – bring letters of the same type or any numbers together.

To **evaluate** means to work out the value of an expression by replacing one or more variables (letters) with numbers.

Example 2.1 illustrates these terms and how to use them.

> **Hint** Look at Chapter 1, pages 1–3, for information on calculating with negative numbers.

N4 **Example 2.1**

a Expand $10(6 - 5p)$.

b Factorise $16m + 24$.

c Simplify $8r + s - 7r - 6s + 12$.

d Evaluate $\frac{2}{3}a + 5b$, given that $a = 12$ and $b = -2$.

a $10(6 - 5p) = 10 \times 6 + 10 \times (-5p)$

$= 60 - 50p$

> Notice the subtraction sign in the brackets.

b $16m + 24 = 8(2m + 3)$

> 8 is the highest common factor of $16m$ and 24.

Check:

$8(2m + 3) = 16m + 24$ ✔

> Check that you have factorised correctly by expanding the brackets again.

c $8r + s - 7r - 6s + 12 = 8r - 7r + s - 6s + 12$

> Collect like terms.

$= r - 5s + 12$

> $s - 6s = 1s - 6s$, but by convention, we do not write 1 in front of a single variable. So do **not** write your final answer as $1r - 5s + 12$.

d $a = 12$ and $b = -2$:

$\frac{2}{3}a + 5b = \frac{2}{3}(12) + 5(-2)$

> Substitute the values given for a and b into the equation.
> $\frac{2}{3}(12)$ means $\frac{2}{3} \times 12$ or $\frac{2}{3}$ of 12.

$= 8 + (-10)$

$= 8 - 10 = -2$

N4 **Exercise 2A**

1 Expand:

a $6(x + 4)$

b $7(a - 11)$

c $2(4 - t)$

d $5(2y - 3)$

e $10(2p + 9q)$

f $-4(2m + 5)$

g $-11(3d - 5)$

h $-(1 - 8z)$

> Hint $-(1 - 8z) = -1(1 - 8z)$

i $6(2a + 3b - 4)$

j $-3(4 - 2x + 3y)$

k $-(a - 2b - c)$

2 Factorise:

a $2x + 12$

b $15c - 5$

c $14p - 21$

d $12a + 15$

e $8x - 40$

f $12y + 30$

g $10 - 8a$

h $100 + 150z$

i $35a + 28b$

j $16p - 32q$

k $25x - 35y + 55z$

l $12 + 20r - 28s$

> Hint Remember to check your answers by expanding your solution to obtain the initial expression.

3 Simplify:

a $5x + 11x$ **b** $8c - 7c$ **c** $12y + y$ **d** $15x - 2x + 11$

e $4t + 8 - 2t$ **f** $8a + 6b + a - 5b$ **g** $5x - 2y - 9x + 6y$ **h** $12p - 4 + 11q - 6p$

i $6a - 7a - 8b + 7b$ **j** $-a + 2b + 3a - 4b$ **k** $1 - x + 2y + 12 - 4x - y$ **l** $6x^2 + 3x - x^2 + x$

4 Evaluate:

a $4x + 12$ when $x = 3$ **b** $10b - 3$ when $b = 5$

c $40 - 6t$ when $t = 7$ **d** $2y + 4$ when $y = -5$

e $\frac{1}{2}w + 5$ when $w = 50$ **f** $100 - \frac{1}{3}p$ when $p = 30$

g $2x + 3y$ given that $x = 9$ and $y = 3$

h $u - 3v$ given that $u = 8$ and $v = -2$

i $50 - 3a - \frac{1}{2}b$ when $a = 10$ and $b = 40$

j $\frac{3}{4}p - \frac{2}{3}q + 4$ when $p = 16$ and $q = 18$

k $x^2 - y^2$ given that $x = 5$ and $y = 1$

> **Hint** Remember that a^2 means $a \times a$.

l $a^2 + b^2$ given that $a = 2$ and $b = -3$

N5 Expanding brackets

This section looks at:

- expanding and simplifying expressions involving more than one pair of brackets
- expanding brackets by multiplying through by a variable (or letter)
- expanding two pairs of brackets by multiplying them together.

> **Hint** Remember that the $\times$ sign is omitted when using brackets. So, for example, $5(4x - 3) - 6(2x - 1)$ means $5 \times (4x - 3) - 6 \times (2x - 1)$ and $(3m + 2)(5m - 3)$ means $(3m + 2) \times (5m - 3)$.

As explained in the previous section, when you expand brackets, you multiply each term inside the brackets separately by whatever is in front of the brackets. Then you simplify the expression obtained by collecting like terms.

When expanding two pairs of brackets, you do the same thing. For example, to expand $(3m + 2)(5m - 3)$, you take the first term in the first pair of brackets, $3m$, and multiply it by the second pair of brackets, $(5m - 3)$. Then you take the second term in the first pair of brackets, $+2$, and multiply it by the second pair of brackets, $(5m - 3)$. Expand each pair of brackets separately before collecting like terms:

$$(3m + 2)(5m - 3) = 3m(5m - 3) + 2(5m - 3)$$
$$= 3m \times 5m + 3m \times (-3) + 2 \times 5m + 2 \times (-3)$$
$$= 15m^2 - 9m + 10m - 6$$
$$= 15m^2 + m - 6$$

N5 **Example 2.2**

Expand:

a $6(3t + 1) - 2(4t - 5)$ **b** $-8x(2x - 1)$ **c** $(2y + 7)(3y - 4)$

a $6(3t + 1) - 2(4t - 5) = 6 \times 3t + 6 \times 1 - 2 \times 4t - 2 \times (-5)$

$$= 18t + 6 - 8t + 10$$
$$= 18t - 8t + 6 + 10 \quad \bullet\!\!-\!\!-\!\!-\!\!-\!\!-\!\!-\!\!-\!\!- \boxed{\text{Collect like terms, then simplify.}}$$
$$= 10t + 16$$

b $-8x(2x - 1) = -8x \times 2x - 8x \times (-1) = -16x^2 + 8x$

c $(2y + 7)(3y - 4) = 2y(3y - 4) + 7(3y - 4)$

$$= 2y \times 3y + 2y \times (-4) + 7 \times 3y + 7 \times (-4)$$
$$= 6y^2 - 8y + 21y - 28$$
$$= 6y^2 + 13y - 28$$

> **Hint** The mnemonic **FOIL** can be used to remember how to expand a pair of brackets. Multiply **F**irst terms, **O**utside, **I**nside and **L**ast.

N5 **Exercise 2B**

1 Expand and simplify:

a $4(x + 5) + 3(x + 1)$ b $8(t + 2) + 4(t - 3)$

c $10(p + 6) + 9(p - 10)$ d $5(y - 4) + 3(y + 5)$

e $6(a - 2) + 2(a - 1)$ f $7(x + 4) - 3(x + 3)$

g $5(p - 1) - 2(p - 8)$ h $7(3 - x) - (4 + x)$

> **Hint** $7(3 - x) - (4 + x) = 7(3 - x) - 1(4 + x)$

★ **2** Expand and simplify:

a $3(2m + 1) + 4(5m + 2)$ b $4(3x + 4) + 6(2x - 3)$

c $6(5 + 2p) - 3(1 + 3p)$ d $7(2y + 1) - 2(4y - 3)$

e $2(2n - 3) - 5(1 - n)$ f $-5(6a - 1) + 4(2a + 3)$

★ **3** Expand:

a $x(x + 5)$ b $y(y - 2)$ c $p(p - 1)$ d $t(6 + t)$

e $w(1 - w)$ f $6x(x + 2)$ g $4r(r - 3)$ h $3l(2 - l)$

i $z(5z + 10)$ j $b(7b - 3)$ k $k(18 - 7k)$ l $3a(2a + 5)$

m $2p(2p - 1)$ n $9d(2 - 3d)$ o $-4x(7x + 4)$ p $-5s(1 - 2s)$

4 Expand the brackets and collect like terms.

a $(x + 5)(x + 2)$ **b** $(t + 1)(t + 8)$

c $(y + 10)(y - 3)$ **d** $(a + 5)(a - 7)$

e $(p - 1)(p + 6)$ **f** $(m - 4)(m + 3)$

g $(q - 2)(q - 10)$ **h** $(x - 4)^2$ | Hint | $(x - 4)^2 = (x - 4)(x - 4)$

 5 Expand the brackets and collect like terms.

a $(2x + 5)(x + 2)$ **b** $(5a + 7)(2a + 3)$ **c** $(3y + 4)(6y + 1)$

d $(5p + 2)^2$ **e** $(7e + 8)(e - 1)$ **f** $(4x + 3)(2x - 3)$

g $(y - 1)(9y + 11)$ **h** $(7b - 4)(3b + 1)$ **i** $(2x - 5)(2x - 3)$

j $(4y - 3)(5y - 1)$ **k** $(3f - 4)^2$ **l** $(1 - 2t)^2$

N5 Factorising

This section will show you how to:

- factorise by removing the highest common factor (a number and/or a variable)
- factorise by recognising a difference of two squares
- factorise a trinomial expression.

▶N5 Factorising by removing the highest common factor

The highest common factor of two or more terms is the greatest factor common to all the terms. It may be a variable (letter), a number, or a combination of the two.

N5 Example 2.3

Factorise fully:

a $8ab - 5ac$ **b** $x^2 + 7x$ **c** $6xy - 9y^2$

a $8ab - 5ac = a(8b - 5c)$ ●————— a is the highest common factor of $8ab$ and $-5ac$.

   Check:

   $a(8b - 5c) = 8ab - 5ac$ ✓ ●————— Check that you have factorised correctly by multiplying the brackets back out again.

b $x^2 + 7x = x(x + 7)$

c $6xy - 9y^2 = 3y(2x - 3y)$ ●————— To **fully** factorise the expression, you must remove the **highest** common factor, which is $3y$.

Exercise 2C

1 Factorise fully:

  a $8m + mn$ b $2ab - 9a$ c $4xy + xz$ d $3pq - 4qr$

  e $12p - 16q$ f $100 + 30x$ g $6uv + 5vw$ h $11ab - ac$

> **Hint**   Remember to check each answer by expanding it to get the initial expression.

2 Factorise fully:

  a $x^2 + 3x$ b $4y^2 - y$ c $6a - a^2$ d $t^2 + t$

  e $6a^2 - 3$ f $5t + 4t^2$ g $12y^2 - 11xy$ h $27a^2 + 18b^2$

 3 Factorise fully:

  a $2x^2 + 4x$ b $6y - 6y^2$ c $25u^2 + 35uv$ d $12ab - 8b^2$

  e $60p^2 + 24pq$ f $81m - 99m^2$ g $70b^2 - 65ab$ h $1600yz + 1200y^2$

Factorising by recognising a difference of two squares

Expand: $(x + y)(x - y)$ and $(a + b)(a - b)$:

$$(x + y)(x - y) = x(x - y) + y(x - y)$$
$$= x^2 - xy + yx - y^2$$
$$= x^2 - xy + xy - y^2$$
$$= x^2 - y^2$$

$$(a + b)(a - b) = a(a - b) + b(a - b)$$
$$= a^2 - ab + ba - b^2$$
$$= a^2 - ab + ab - b^2$$
$$= a^2 - b^2$$

So:

$$(x + y)(x - y) = x^2 - y^2 \qquad \text{and} \qquad (a + b)(a - b) = a^2 - b^2$$

Conversely:

$$x^2 - y^2 = (x + y)(x - y) \qquad \text{and} \qquad a^2 - b^2 = (a + b)(a - b)$$

Both $x^2 - y^2$ and $a^2 - b^2$ express a difference of two squares, because they both have the form $(something)^2 - (something\ else)^2$.

> **Important**
>
> The rule for factorising a **difference of two squares** is:
>
> $$p^2 - q^2 = (p + q)(p - q) \text{ or } (p - q)(p + q)$$

> **Hint**   When factorising an expression that is the difference of two squares, one bracket should contain a '+' sign and the other should contain a '−', but the order is not important.

Example 2.4 shows how to factorise such expressions.

17

N5 **Example 2.4**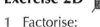

Factorise:

a $m^2 - n^2$　　　　b $x^2 - 100$　　　　c $144p^2 - 49$　　　d $1 - 9t^2$

a $m^2 - n^2 = (m + n)(m - n)$

b $x^2 - 100 = x^2 - 10^2$

$\quad\quad\quad\quad = (x + 10)(x - 10)$

c $144p^2 - 49 = (12p)^2 - 7^2$ ●——————————————$\boxed{144 = 12^2, \text{ so } 144p^2 = (12p)^2}$

$\quad\quad\quad\quad = (12p + 7)(12p - 7)$

d $1 - 9t^2 = 1^2 - (3t)^2$ ●——————————————$\boxed{1 = 1^2 \text{ and } 9 = 3^2, \text{ so } 9t^2 = (3t)^2}$

$\quad\quad\quad = (1 + 3t)(1 - 3t)$

N5 **Exercise 2D**

★ **1** Factorise:

    a $c^2 - d^2$　　　　b $u^2 - v^2$　　　　c $y^2 - x^2$　　　　d $x^2 - 81$

    e $a^2 - 4$　　　　f $p^2 - 36$　　　　g $25 - y^2$　　　　h $64 - x^2$

2 Factorise:

    a $4x^2 - 9$　　　b $16m^2 - 1$　　　c $4 - 81p^2$　　　d $121 - 100a^2$

    e $25a^2 - 36b^2$　f $4p^2 - 9q^2$　　　g $169 - 400n^2$　　h $4x^2 - \frac{1}{4}y^2$

N5 **Factorising trinomials**

A **trinomial** expression has three parts (*tri-* means 'three'). To factorise a trinomial expression such as $y^2 + 9y + 20$, follow these steps:

- Look at the first term, y^2, and find its factors: y and y.

  Set up the factorisation: $y^2 + 9y + 20 = (y \,...)(y \,...)$

- Look at the last term, 20, and find its factor pairs. These could be 1 and 20, 2 and 10 or 4 and 5.

- Look at the middle term, $9y$. Of the possible factor pairs of 20, which pair of numbers add to give 9?

    $1 + 20 = 21$ ✗　　　　　$2 + 10 = 12$ ✗　　　　　$4 + 5 = 9$ ✓

Complete the factorisation: $y^2 + 9y + 20 = (y + 4)(y + 5)$

Always check your answer by expanding it, to check you obtain the initial expression.

N5 **Example 2.5**

Factorise:

a $x^2 + 8x + 7$ b $t^2 + 5t + 6$ c $p^2 + 12p + 20$

a $x^2 + 8x + 7 = (x + ...)(x + ...)$

$\qquad = (x + 1)(x + 7)$ ●————— Look for factor pairs of 7 that add to 8; $1 \times 7 = 7, 1 + 7 = 8$ ✔

Check: ●————— Check that you have factorised correctly by expanding the brackets.

$(x + 1)(x + 7) = x(x + 7) + 1(x + 7)$

$\qquad = x^2 + 7x + x + 7$

$\qquad = x^2 + 8x + 7$ ✔

b $t^2 + 5t + 6 = (t + 2)(t + 3)$ ●————— Look for factor pairs of 6 that add to 5; $2 \times 3 = 6, 2 + 3 = 5$ ✔

Check:

$(t + 2)(t + 3) = t(t + 3) + 2(t + 3)$

$\qquad = t^2 + 3t + 2t + 6$

$\qquad = t^2 + 5t + 6$ ✔

c $p^2 + 12p + 20 = (p + 2)(p + 10)$ ●————— Look for factor pairs of 20 that add to 12; $2 \times 10 = 20, 2 + 10 = 12$ ✔

Check:

$(p + 2)(p + 10) = p(p + 10) + 2(p + 10)$

$\qquad = p^2 + 10p + 2p + 20$

$\qquad = p^2 + 12p + 20$ ✔

N5 **Exercise 2E**

1 Factorise:

a $x^2 + 6x + 5$ b $a^2 + 12a + 11$ c $m^2 + 4m + 3$

d $y^2 + 9y + 14$ e $t^2 + 8t + 15$ f $x^2 + 23x + 22$

> **Hint** Remember that 1 is a factor of any number.

★ 2 Factorise:

a $x^2 + 7x + 12$ b $k^2 + 8k + 12$ c $w^2 + 12w + 20$

d $m^2 + 17m + 16$ e $t^2 + 11t + 24$ f $p^2 + 6p + 9$

3 Factorise:

a $a^2 + 29a + 100$ b $p^2 + 14p + 13$ c $v^2 + 15v + 44$

d $x^2 + 40x + 400$ e $n^2 + 50n + 49$ f $t^2 + 2t + 1$

N5 **Factorising trinomials containing negative terms**

The same method is used to factorise a trinomial containing one or more negative signs, for example, $x^2 - 7x - 8$:

- Look at the first term, x^2, and find its factors: x and x.

  Set up the factorisation: $x^2 - 7x - 8 = (x \ldots)(x \ldots)$

- Look at the last term, -8, and find its factor pairs. These could be 1 and -8, -1 and 8, 2 and -4 or -2 and 4.

- Look at the middle term, $-7x$. Of the possible factor pairs of -8, which pair of numbers add to give -7?

$$1 + -8 = -7 \ \checkmark \qquad -1 + 8 = 7 \ \times \qquad 2 + -4 = -2 \ \times \qquad -2 + 4 = 2 \ \times$$

Complete the factorisation, making sure you put the minus sign in the correct place:
$x^2 - 7x - 8 = (x + 1)(x - 8)$

Again, always check your answer by expanding it, to check you obtain the initial expression.

> **Hint** Look at Chapter 1, page 2, for the rules on multiplying negative numbers.

N5 **Example 2.6** 🖩

Factorise:

a $t^2 + 10t - 11$ **b** $q^2 - 8q + 12$ **c** $m^2 - m - 20$

a $t^2 + 10t - 11 = (t - 1)(t + 11)$

> Look for factor pairs of -11 that add to 10; $-1 \times 11 = -11$, $-1 + 11 = 10$ ✔

Check:

> Check your answer.

$(t - 1)(t + 11) = t(t + 11) - 1(t + 11)$

$\qquad = t^2 + 11t - t - 11$

$\qquad = t^2 + 10t - 11 \ \checkmark$

b $q^2 - 8q + 12 = (q - 2)(q - 6)$

> Look for factor pairs of 12 that add to -8; $-2 \times -6 = 12$, $-2 + -6 = -8$ ✔

Check:

$(q - 2)(q - 6) = q(q - 6) - 2(q - 6)$

$\qquad = q^2 - 6q - 2q + 12$

$\qquad = q^2 - 8q + 12 \ \checkmark$

c $m^2 - m - 20 = (m + 4)(m - 5)$

> Look for factor pairs of -20 that add to -1; $4 \times -5 = -20$, $4 + -5 = -1$ ✔

Check:

$(m + 4)(m - 5) = m(m - 5) + 4(m - 5)$

$\qquad = m^2 - 5m + 4m - 20$

$\qquad = m^2 - m - 20 \ \checkmark$

Exercise 2F

1 Factorise:

a $p^2 - 4p + 3$ b $x^2 + x - 2$ c $x^2 - 6x - 7$

d $t^2 + 4t - 5$ e $y^2 - 2y + 1$ f $m^2 - 12m - 13$

2 Factorise:

a $k^2 + k - 12$ b $p^2 - 12p + 20$ c $y^2 - 15y - 16$

d $a^2 - 8a + 15$ e $w^2 + 3w - 10$ f $x^2 - 2x - 35$

3 Factorise:

a $x^2 - 40x + 400$ b $y^2 + 3y + 2$ c $t^2 - 99t - 100$

d $p^2 + 24p + 144$ e $r^2 - 3r - 10$ f $h^2 - 10h + 25$

 **Algebraic fractions**

An algebraic fraction is any fraction involving a variable, whether in the numerator or the denominator, for example, $\frac{1}{a}$, $\frac{a+b}{5}$ and $\frac{x^2-6}{y}$

This section shows how to:

- simplify algebraic fractions
- add, subtract, multiply or divide using algebraic fractions.

Simplifying algebraic fractions

A fraction is a way of representing division: $\frac{2}{3}$ is the same as $2 \div 3$. You can **simplify** an

algebraic fraction by dividing the numerator and the denominator by a common factor of both. When you simplify fully by dividing by the highest common factor, you reduce the fraction to its simplest form Examples 2.7 and 2.8 show how to do this.

> Hint | Remember that 'anything divided by itself' equals 1.

Example 2.7

Simplify:

a $\dfrac{x^2}{x}$ b $\dfrac{y^2}{y^4}$ c $\dfrac{8p^2q}{6pq^3}$

a $\dfrac{x^2}{x} = \dfrac{x \times x^1}{x_1} = \dfrac{x}{1} = x$ •————— x is the highest common factor of the numerator and the denominator.

b $\dfrac{y^2}{y^4} = \dfrac{{}^1y \times y^1}{{}_1y \times y_1 \times y \times y} = \dfrac{1}{y^2}$

c $\dfrac{8p^2q}{6pq^3} = \dfrac{{}^48 \times p^1 \times p \times q^1}{{}_36 \times {}_1p \times {}_1q \times q \times q} = \dfrac{4p}{3q^2}$ •——— Be careful; cancel one term at a time.

N5 **Example 2.8** 🖩

Simplify:

a $\dfrac{(a+4)}{(a-3)(a+4)}$

b $\dfrac{(p+1)(p-1)^3}{(p+1)^2(p-1)}$

c $\dfrac{2(1-x)(x+6)^2}{4(1-x)^2(x+6)}$

a $\dfrac{(a+4)}{(a-3)(a+4)} = \dfrac{^1(a+4)}{(a-3)(a+4)_1} = \dfrac{1}{(a-3)}$ or $\dfrac{1}{a-3}$

b $\dfrac{(p+1)(p-1)^3}{(p+1)^2(p-1)} = \dfrac{^1(p+1)\,^1(p-1)(p-1)(p-1)}{_1(p+1)(p+1)(p-1)_1}$

$= \dfrac{(p-1)^2}{(p+1)}$ or $\dfrac{(p-1)^2}{p+1}$

c $\dfrac{2(1-x)(x+6)^2}{4(1-x)^2(x+6)} = \dfrac{^1 2(1-x)\,^1(x+6)(x+6)}{_2 4(1-x)(1-x)(x+6)_1}$

$= \dfrac{(x+6)}{2(1-x)}$ or $\dfrac{x+6}{2(1-x)}$

N5 **Exercise 2G** 🖩

1 Simplify:

a $\dfrac{a^2}{a}$

b $\dfrac{x}{x^2}$

c $\dfrac{p^5}{p^2}$

d $\dfrac{q^3}{q^5}$

e $\dfrac{8x^2}{4x}$

f $\dfrac{3y}{9y^3}$

g $\dfrac{25m^2}{35m^2}$

h $\dfrac{7t^6}{7t^5}$

★ 2 Reduce each expression to its simplest form.

a $\dfrac{12ab}{4a}$

b $\dfrac{5y}{10xy}$

c $\dfrac{14uv}{14uv}$

d $\dfrac{32abc}{24acd}$

e $\dfrac{x^2y}{xy^2}$

f $\dfrac{10p^2q^2}{5pq^2}$

g $\dfrac{12ab}{16a^2b^2}$

h $\dfrac{25u^2v^3}{75uv}$

3 Simplify:

a $\dfrac{(x+1)(x+5)}{(x+5)}$

b $\dfrac{(y-3)}{(y-3)(y+2)}$

c $\dfrac{6(t-1)(t+1)}{2(t-2)(t+1)}$

d $\dfrac{5(q-4)(q-1)^2}{(q-4)(q-1)}$

e $\dfrac{4x^2(x+1)}{8x(x+1)^2}$

f $\dfrac{(a+1)^2(a+4)^3}{(a+1)^2(a+4)}$

4 Simplify:

a $\dfrac{(4-x)^2(x+3)}{(4-x)(x+3)^2}$

b $\dfrac{12(y+1)^2(y+3)^2}{2(y+1)(y+3)}$

c $\dfrac{5(3-t)(t+4)^2}{30(3-t)^2(t+4)}$

N5 Calculating with algebraic fractions

In order to add or subtract algebraic fractions, the rules are the same as for adding and subtracting numerical fractions: the fractions must have a common denominator (that is, the expressions on the bottom of the fractions must be the same).

If the algebraic fractions do not have a common denominator, you need to convert one or both of them so they do have a common denominator. To identify a common denominator you can multiply the denominators. In order that you don't change the value of the fraction you must multiply the numerator by the same number or variable as you multiply the denominator.

To multiply two fractions together, you simply multiply the two numerators together and the two denominators together. Then check to see if the answer can be simplified.

To divide one fraction by another, turn the second fraction upside down (that is, invert it) and then multiply.

Example 2.9 shows how to apply these methods and use the four operations $(+, -, \times$ and $\div)$ to calculate with fractions.

> **Hint** Look at Chapter 1, pages 4–8, for further help with calculating with numerical fractions.

N5 Example 2.9

Find:

a $\dfrac{x}{4} + \dfrac{3}{x}$ b $\dfrac{5}{a} - \dfrac{2}{b}$ c $\dfrac{t+1}{2} \times \dfrac{8}{(t+1)^2}$ d $\dfrac{ab^2}{5} \div \dfrac{a^2 b}{15}$

a $\dfrac{x}{4} + \dfrac{3}{x} = \dfrac{x \times x}{4 \times x} + \dfrac{3 \times 4}{x \times 4}$ Multiply $\dfrac{x}{4}$ by $\dfrac{x}{x}$ and multiply $\dfrac{3}{x}$ by $\dfrac{4}{4}$

$= \dfrac{x^2}{4x} + \dfrac{12}{4x} = \dfrac{x^2 + 12}{4x}$

b $\dfrac{5}{a} - \dfrac{2}{b} = \dfrac{5b}{ab} - \dfrac{2a}{ab} = \dfrac{5b - 2a}{ab}$ Multiply $\dfrac{5}{a}$ by $\dfrac{b}{b}$ and multiply $\dfrac{2}{b}$ by $\dfrac{a}{a}$

c $\dfrac{t+1}{2} \times \dfrac{8}{(t+1)^2} = \dfrac{{}^4 8(t+1)}{{}_1 2(t+1)(t+1)} = \dfrac{4}{t+1}$

d $\dfrac{ab^2}{5} \div \dfrac{a^2 b}{15} = \dfrac{ab^2}{5} \times \dfrac{15}{a^2 b} = \dfrac{{}^3 15 \times a \times b \times b}{{}_1 5 \times a \times a \times b} = \dfrac{3b}{a}$

N5 Exercise 2H

★ 1 Find:

a $\dfrac{x}{2} + \dfrac{x}{5}$ b $\dfrac{t}{3} - \dfrac{t}{4}$ c $\dfrac{p}{3} \times \dfrac{p}{2}$ d $\dfrac{a}{5} \div \dfrac{a}{10}$

e $\dfrac{y}{3} + \dfrac{7}{y}$ f $\dfrac{w}{8} - \dfrac{2}{w}$ g $\dfrac{u}{6} \times \dfrac{18}{u}$ h $\dfrac{x}{2} \div \dfrac{5}{x}$

2 Find:

a $\dfrac{7}{a} + \dfrac{2}{b}$ b $\dfrac{6}{u} - \dfrac{3}{v}$ c $\dfrac{c}{5} \times \dfrac{d}{3}$ d $\dfrac{p}{14} \div \dfrac{q}{7}$

e $\dfrac{(x+5)^2}{2} \times \dfrac{8}{x+5}$ f $\dfrac{3}{2-t} \times \dfrac{(2-t)^2}{9}$ g $\dfrac{x^2 y}{15} \div \dfrac{xy}{3}$ h $\dfrac{21}{pq^2} \div \dfrac{14}{p^2 q}$

N5 **Chapter 2 review**

1 Expand the brackets and simplify by collecting like terms.

   a $5(4x + 2) - 3(2x - 1)$ **b** $-4t(3t + 5)$ **c** $(4m - 3)(5m + 7)$

2 Factorise:

   a $3pq + 2qr$ **b** $8m - m^2$ **c** $15t^2 + 25st$

3 Factorise:

   a $u^2 - v^2$ **b** $t^2 - 81$ **c** $25a^2 - 36$ **d** $16u^2 - 1$

4 Factorise:

   a $x^2 + 6x + 5$ **b** $y^2 + 6y + 8$ **c** $a^2 + 14a + 40$

5 Factorise:

   a $p^2 + 6p - 7$ **b** $x^2 - 9x + 20$ **c** $a^2 - 2a - 15$

6 Simplify:

   a $\dfrac{p^3}{p}$ **b** $\dfrac{a^2}{a^5}$ **c** $\dfrac{25ab^2}{35a^3b}$

7 Simplify:

   a $\dfrac{(t - 2)(t + 6)}{(t + 6)}$ **b** $\dfrac{(m + 2)^2 (m + 5)}{(m + 2)(m + 5)^3}$ **c** $\dfrac{3(x - 3)(2 - x)^2}{9(x - 3)^2 (2 - x)}$

8 Simplify:

   a $\dfrac{7}{x} + \dfrac{3}{y}$ **b** $\dfrac{a}{6} - \dfrac{5}{a}$ **c** $\dfrac{(y + 2)^2}{15} \times \dfrac{3}{y + 2}$ **d** $\dfrac{u^2v}{16} \div \dfrac{uv^2}{4}$

- I can work with algebraic expressions involving expansion of brackets. ★ Exercise 2B Q2, Q3, Q5
- I can factorise an algebraic expression involving a common factor. ★ Exercise 2C Q3
- I can factorise an algebraic expression involving a difference of two squares. ★ Exercise 2D Q1
- I can factorise an algebraic expression involving a trinomial. ★ Exercise 2E Q2 ★ Exercise 2F Q2
- I can reduce an algebraic fraction to its simplest form. ★ Exercise 2G Q2
- I can apply one of the four operations (+, −, × or ÷) to algebraic fractions. ★ Exercise 2H Q1

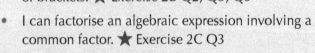

3 Trigonometry

This chapter will show you how to:
- identify the different sides of a right-angled triangle
- use trigonometry to calculate the missing side in a right-angled triangle
- use trigonometry to calculate an angle in a right-angled triangle.

You should already know:
- how to round a number to a given number of decimal places
- that a bearing is measured clockwise from the north line.

N4 Labelling the sides of a right-angled triangle and investigating the trigonometric ratios

Trigonometry can be used to calculate the missing side or angle in a right-angled triangle. If an angle in a right-angled triangle is labelled as $x°$, as shown in the diagram, the sides are labelled as follows:

- **hypotenuse:** this is the side opposite the right angle, and it is always the longest side; it is often shortened to *hyp*

- **opposite:** this is the side directly opposite the given angle $x°$; it is often shortened to *opp*

- **adjacent:** this is the side that touches both the angle $x°$ and the right angle; it is the side 'next to' $x°$ that isn't the hypotenuse, and is often shortened to *adj*.

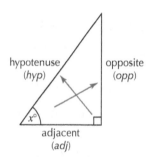

Investigating right-angled triangles part 1: the tangent ratio

- Draw three right-angled triangles with a 35° angle in each of them.
  The triangles must be different sizes but must contain a 35° angle.
  For example, draw triangles with baselines of 5 cm, 7 cm and 9 cm.

> **Hint** Before you start any work on trigonometry, always check that your calculator has a D or DEG at the top of the screen. Your teacher, or the instruction manual, will help you change the settings to get a D or DEG on the screen if you don't have it already.

- Label the sides of the triangles as *hyp*, *adj* and *opp*.

- Measure the lengths of all the sides in the triangles as accurately as you can and write your measurements on each side.

- Divide the length of the opposite by the length of the adjacent for each of the triangles, that is, calculate the ratio $\frac{\text{opposite}}{\text{adjacent}}$ for each of your three triangles.

- What do you notice?

  You should get approximately the same answer for each of the three calculations.

- On your scientific calculator, type in $\tan 35°$.
- You should get $0.700\,207\,538$ correct to 9 decimal places.
- This should be approximately the same as your answers to $\frac{\text{opp}}{\text{adj}}$ for each of your three triangles.

- So $\tan 35° = \frac{\text{opp}}{\text{adj}}$ for the values of the opposite and adjacent sides that you have measured in your triangles.

- Repeat these steps and check that it works when you use an angle of $55°$ (the other angle in your three right-angled triangles).

- You should find that $\tan 55° = \frac{\text{opp}}{\text{adj}} = 1.428\,148\,007$ correct to 9 decimal places for the values of the opposite and adjacent sides that you have measured in your triangles.

> **Important**
>
> The tangent function is given by:
>
> $$\tan x° = \frac{\text{opposite}}{\text{adjacent}} = \frac{\text{opp}}{\text{adj}}$$

> **Hint** tan is short for tangent.

N4 Using the tangent ratio to calculate the length of a missing side in a right-angled triangle

You can use the rule $\tan x° = \frac{\text{opp}}{\text{adj}}$ to calculate the length of either the opposite or the adjacent sides in a right-angled triangle if you know the length of the other side and an angle.

N4 Example 3.1

Calculate the length of the side marked a in the triangle.

Give your answer to 1 decimal place (1 d.p.).

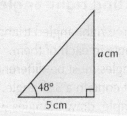

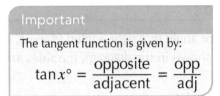

Label the sides of the triangle hyp, adj and opp.

$$\tan x° = \frac{\text{opp}}{\text{adj}}$$

Write down the formula for tan.

$$\tan 48° = \frac{a}{5}$$ •────── Substitute values for the angle = 48°, opp = a and adj = 5.

$$5 \times \tan 48° = a$$ •────── Multiply both sides of the equation by 5 in order to get a on its own.

$$a = 5·553\,062\,57...$$

$$= 5·6\,cm\,(1\,d.p.)$$ •────── Calculate a, writing down both your unrounded and rounded answers.

N4 **Example 3.2** 🖩

Calculate the length of the side marked b in the triangle.

Give your answer to 1 decimal place (1 d.p.).

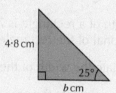

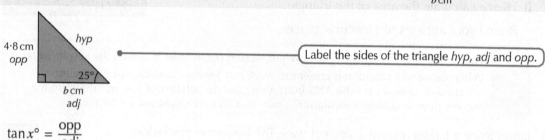

$$\tan x° = \frac{opp}{adj}$$

Label the sides of the triangle hyp, adj and opp.

$$\tan 25° = \frac{4·8}{b}$$ •────── Substitute values for the angle = 25°, opp = 4.8 and adj = b.

$$b \times \tan 25° = 4·8$$ •────── Multiply both sides of the equation by b.

$$b = \frac{4·8}{\tan 25°}$$ •────── Divide both sides of the equation by tan 25°.

$$= 10·293\,633\,22...$$

$$= 10·3\,cm\,(1\,d.p.)$$ •────── Remember to write down both your unrounded and rounded answers.

N4 **Exercise 3A** 🖩

★ 1 Calculate the length of the side marked with a letter in each triangle. Round your
answers to 1 decimal place.

a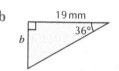

b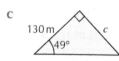

c

d

e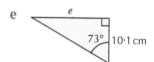

f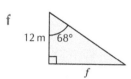

Hint Remember to give the units in your answer.

★ **2** Calculate the length of the side of the side marked with a letter in each triangle. Round your answers to 1 decimal place.

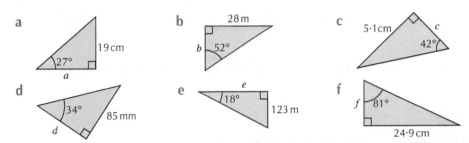

a   19 cm   27°   a

b   28 m   b   52°

c   5·1 cm   42°   c

d   34°   85 mm   d

e   e   18°   123 m

f   f   81°   24·9 cm

3 The length of a rectangle is 7·8 centimetres. The angle between the diagonal of the rectangle and its length is 30°.

  a Calculate the width of the rectangle.

  b Hence calculate the area of the triangle.

    Round your answers to 1 decimal place.

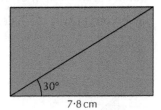

30°

7·8 cm

> **Hint** **Hence** means use the answer you have just worked out in order to answer the new question.
> When using one calculated answer to work out another answer, always use the unrounded answer (use the ANS button on your calculator). If you use the rounded answer in your second calculation, your final answer might not be correct.

4 James leans a ladder against a vertical wall. The bottom of the ladder is 1·3 metres horizontally from the bottom of the wall and the ladder makes an angle of 57° with the ground.

Calculate how far up the wall the top of the ladder reaches. Round your answer to 1 decimal place.

57°

1·3 m

★ **5** A ladder stands on the ground and leans on a vertical wall. It reaches 7·2 metres up the wall and makes an angle of 70° with the ground.

Calculate the distance of the base of the ladder from the bottom of the wall. Round your answer to 1 decimal place.

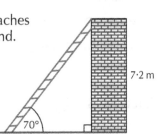

7·2 m

70°

6 A ship leaves a harbour and sails on a bearing of 040°. When the ship stops, it is 50 km north of its starting point.

How far east of its starting point is the ship? Round your answer to 1 decimal place.

> **Hint** Sketch and label a diagram to show the situation.

7 a Calculate the length of the base, *b*, in this isosceles triangle.

  b Hence calculate the area of the isosceles triangle.

  Round your answers to 1 decimal place.

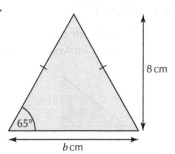

N4 Using the tangent ratio to calculate an angle in a right-angled triangle

You can use the rule $\tan x° = \dfrac{\text{opp}}{\text{adj}}$ to calculate the size of an angle in a right-angled triangle if you know the lengths of the opposite and adjacent sides.

On your scientific calculator you will need to use the **tan⁻¹** function. It is usually above the tan button and if you press SHIFT and then tan, tan⁻¹ will appear on the calculator display.

N4 Example 3.3 🖩

Calculate the size of the angle $\theta°$ in this triangle.

Give your answer to 2 decimal places (2 d.p.).

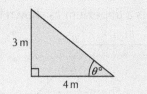

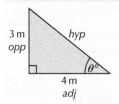

$\tan x° = \dfrac{\text{opp}}{\text{adj}}$

$\tan\theta° = \dfrac{3}{4}$ ⸺ Substitute values for the angle = $\theta°$, *opp* = 3 and *adj* = 4.

$\theta° = \tan^{-1}\left(\dfrac{3}{4}\right)$ ⸺ Use tan⁻¹ on your calculator.

Hint If your calculator does not have a fraction button on it, you can do $\theta° = \tan^{-1}(3 \div 4)$.

$\theta° = 36{\cdot}869\,897\,65...$

$= 36{\cdot}87°$ (2 d.p.) ⸺ Write down both your unrounded and rounded answers.

N4 **Exercise 3B** 🖩

★ 1 Calculate the size of the angle marked with θ or a letter in each right-angled triangle.
Give your answers to 2 decimal places.

a

5 cm
$\theta°$
12 cm

b

1·9 m
$\theta°$
2·8 m

c

73 mm
$\theta°$
22 mm

d

1·5 cm
$a°$
9·5 cm

e

18 cm
23 cm
$b°$

f

10 cm
$c°$
8 cm
6 cm

2 A rectangle has sides of length 8 centimetres and
5 centimetres.

Calculate the angle between the longer side of the
rectangle and a diagonal. Give your answer to 2 decimal
places.

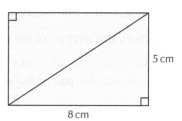

5 cm

8 cm

3 *ABCD* is a trapezium as shown below.

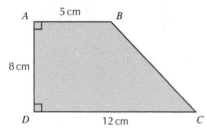

A   5 cm   B

8 cm

D   12 cm   C

Calculate the size of angle *BCD*. Give your answer to 2 decimal places.

★ 4 An isosceles triangle has a base of 10 centimetres and a perpendicular height of
7 centimetres.

Calculate the size of all three angles in the triangle. Give your answer to 2 decimal
places.

5 St Andrews is 54 kilometres to the north of Glasgow and 91 kilometres to the east.

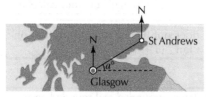

N
N   St Andrews
$a°$
Glasgow

a Calculate the angle marked $a°$.

b What is the bearing of St Andrews from Glasgow?

c What is the bearing of Glasgow from St Andrews?

Give your answers to the nearest whole number.

N4 ## Using the sine ratio to calculate the length of a missing side and the size of an angle in a right-angled triangle

Investigating right-angled triangles part 2: the sine ratio

Look back at the investigation you did at the start of this chapter (pages 25–26).

In the investigation, you divided the length of the opposite side by the length of the adjacent side for each of the triangles, that is, $\dfrac{\text{opp}}{\text{adj}}$

Using the same measurements, this time divide the length of the opposite side by the length of the hypotenuse for each of the triangles, that is, calculate the ratio $\dfrac{\text{opposite}}{\text{hypotenuse}}$ for each of your three triangles.

Now work out $\sin 35°$ on your calculator. You should find this is approximately the same as your answers.

> **Important**
>
> The sine function is given by:
>
> $$\sin x° = \frac{\text{opposite}}{\text{hypotenuse}} = \frac{\text{opp}}{\text{hyp}}$$

> **Hint** sin is short for sine.

N4 ### Example 3.4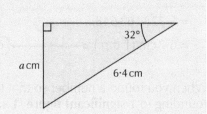

Calculate the length of the side marked a in the triangle.

Give your answer to 2 decimal places (2 d.p.).

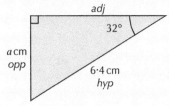

$$\sin x° = \frac{\text{opp}}{\text{hyp}}$$ ⟵ Write down the formula for sine.

$$\sin 32° = \frac{a}{6·4}$$ ⟵ Substitute values for the angle = 32°, $opp = a$ and $hyp = 6.4$

$$6·4 \times \sin 32° = a$$

$$a = 3·391\,483\,291...$$

$$= 3·39\,\text{cm (2 d.p.)}$$ ⟵ Calculate a, writing down both your unrounded and rounded answers.

N4

Example 3.5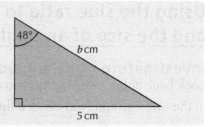

Calculate the length of the side marked *b* in the triangle.

Give your answer to 1 decimal place (1 d.p.).

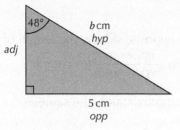

$$\sin x° = \frac{\text{opp}}{\text{hyp}}$$

$$\sin 48° = \frac{5}{b}$$ ⟵ Substitute values.

$$b \times \sin 48° = 5$$

$$b = \frac{5}{\sin 48°}$$

$$= 6{\cdot}728\,163\,648...$$

$$= 6{\cdot}7\,\text{cm (1 d.p.)}$$ ⟵ Calculate *b*.

N5 When you round a number so that there is only **1 non-zero digit in the answer**, it is called rounding to **1 significant figure** (**1 s.f.** or **1 sig. fig.**).

When you are asked to give your answer to 1 s.f. you need:

* to find the first non-zero digit
* to decide the place value of the digit
* to round to that place value using the usual rules for rounding.

This can be extended to 2, 3 or more significant figures.

For example:

* 369.5 is 400 (1 s.f.)
* 2.896 is 2.9 (2 s.f.)
* 7.27434 is 7.27 (3 s.f.)

N4 **Example 3.6** 🖩

Calculate the size of the angle $\theta°$ in this triangle.

Give your answer to 3 significant figures (3 s.f.).

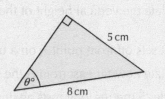

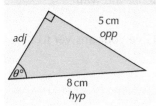

Hint   Rounding to a given number of significant figures is N5 work.

$$\sin x° = \frac{\text{opp}}{\text{hyp}}$$

$$\sin \theta° = \frac{5}{8}$$ ●————(Substitute values.)

$$\theta° = \sin^{-1}\left(\frac{5}{8}\right)$$ ●————(Use $\sin^{-1}$ on your calculator to calculate $\theta°$.)

Hint   If your calculator does not have a fraction button on it, you can do $\theta° = \sin^{-1}(5 \div 8)$.

$$\theta° = 38{\cdot}682\,187\,45...$$

$$= 38{\cdot}7° \text{ (3 s.f.)}$$

N4 **Exercise 3C** 🖩

 1   Calculate the length of the side marked with a letter in each triangle.
Round your answers to 3 significant figures, if required.

a

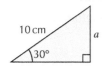

b

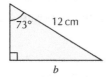

c

d

e

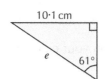

f

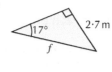

Hint   Rounding to a given number of significant figures is N5 work.

★ 2   Calculate the size of the angle marked with a letter in each right-angled triangle.
Give your answers to 1 decimal place.

a

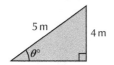

b

c

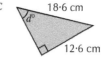

★ 3 A kite on a 23·2 m string is flown at an 81° angle to the ground.

Calculate the vertical height of the kite above the ground. Give your answer to 1 decimal place.

4 A boat sets off from point A on a bearing of 060° and travels for 13·5 km.

Calculate how far east from A the boat is. Give your answer to 1 decimal place.

★ 5 A ladder 5 metres long rests against a vertical wall. The ladder touches the wall 3·8 metres above the ground.

Find the angle that the ladder makes with the horizontal ground. Give your answer to 1 decimal place.

6 The diagram shows the side of a shed.

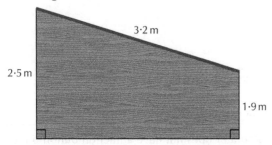

Calculate the size of the angle between the roof and the horizontal. Give your answer to 1 decimal place.

N4 Using the cosine ratio to calculate the length of a missing side and the size of an angle in a right-angled triangle

Investigating right-angled triangles part 3: the cosine ratio

Look back again at the investigation you did at the start of this chapter (pages 25–26).

In the investigation, you divided the length of the opposite side by the length of the adjacent side for each of the triangles, that is, $\frac{\text{opp}}{\text{adj}}$

Using the same measurements, this time divide the length of the adjacent side by the length of the hypotenuse for each of the triangles, that is, calculate the ratio $\frac{\text{adjacent}}{\text{hypotenuse}}$ for each of your three triangles. Now work out $\cos 35°$ on your calculator. This should be approximately the same as your answers.

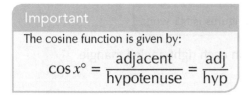

> **Important**
> The cosine function is given by:
> $$\cos x° = \frac{\text{adjacent}}{\text{hypotenuse}} = \frac{\text{adj}}{\text{hyp}}$$

Hint   cos is short for cosine.

N4 **Example 3.7** ▦

Calculate the length of the side marked *a* in the triangle.

Give your answer to 3 significant figures (3 s.f.).

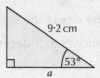

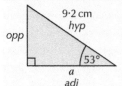

$$\cos x° = \frac{adj}{hyp}$$ ●————————————————(Write down the formula for cos.)

$$\cos 53° = \frac{a}{9·2}$$ ●————————————————(Substitute values.)

$$9·2 × \cos 53° = a$$

$$a = 5·536\,698\,213...$$

$$= 5·54 \text{ cm (3 s.f.)}$$ ●———(Calculate *a*, writing down both your unrounded and rounded answers.)

N4 **Example 3.8** ▦

Calculate the length of the side marked *b* in the triangle.

Give your answer to 1 decimal place (1 d.p.).

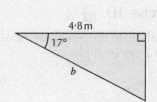

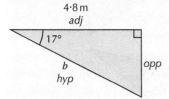

$$\cos x° = \frac{adj}{hyp}$$

$$\cos 17° = \frac{4·8}{b}$$ ●————————————————(Substitute values.)

$$b × \cos 17° = 4·8$$

$$b = \frac{4·8}{\cos 17°}$$

$$= 5·019\,320\,431...$$

$$= 5·0 \text{ m (1 d.p.)}$$ ●———(Calculate *b*.
You need to write the zero after the decimal point because the question specifies an answer to 1 d.p.)

N4 **Example 3.9**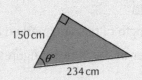

Calculate the size of the angle $\theta°$ in this triangle.

Give your answer to 2 decimal places (2 d.p.).

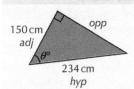

150 cm
adj

opp

234 cm
hyp

$$\cos x° = \frac{adj}{hyp}$$

$$\cos \theta° = \frac{150}{234}$$

$$\theta° = \cos^{-1}\left(\frac{150}{234}\right)$$

Use $\cos^{-1}$ on your calculator to calculate $\theta°$.

Hint   If your calculator does not have a fraction button on it, you can do $\theta = \cos^{-1}(150 \div 234)$.

$$= 50{\cdot}131\,658\,45...$$

$$= 50{\cdot}13° \text{ (2 d.p.)}$$

N4 **Exercise 3D**

★ 1 Calculate the length of the side marked with a letter in each triangle. Give your answers to 2 decimal places.

a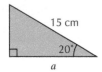

15 cm

20°

a

b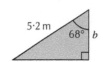

5·2 m

68° b

c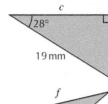

c

28°

19 mm

d

1.4 cm

31°

d

e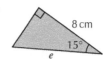

8 cm

15°

e

f

f

81°

9·1 m

★ 2 Calculate the size of the angle marked with a letter or θ in each right-angled triangle. Give your answers to 1 decimal place.

a

17 cm

$\theta°$

10 cm

b

3·75 m

7·5 m $\theta°$

c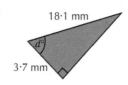

18·1 mm

$d°$

3·7 mm

3 A walker sets off from a point B on a bearing of 052° and walks for 7 kilometres.

Calculate how far north from B the walker is. Give your answer to 3 significant figures.

4 *ABCD* is a kite.

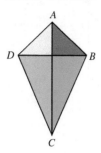

> **Hint** Sketch a diagram and add in all the information you know. The diagonals of a kite meet at 90°.

Angle *ADB* is 30° and *AB* is 12 centimetres.

Calculate the length of *DB*. Give your answer to 3 significant figures.

5 A building regulation states that 8° is the maximum angle between the slope of a ramp and the ground.

A new ramp has a slope of 9 metres and the end of the ramp is 8.8 metres from the building.

a Does the new ramp meet the building regulation? Justify your answer.

b What is the minimum distance from the building to the start of the ramp that would meet the building regulation? Give your answer to 3 significant figures.

N4 Deciding which trigonometric ratio to use: SOH CAH TOA

The three trigonometric ratios are:

$$\sin x° = \frac{opp}{hyp} \qquad \cos x° = \frac{adj}{hyp} \qquad \tan x° = \frac{opp}{adj}$$

You can remember them using the term SOH CAH TOA:

$$\sin x° = \frac{opp}{hyp} \qquad \cos x° = \frac{adj}{hyp} \qquad \tan x° = \frac{opp}{adj}$$

When you need to decide which ratio to use, follow these steps.

1 Label the sides of your right-angled triangle.

2 Decide which sides you are given information about (or you need to calculate) and tick those sides in SOH CAH TOA.

3 Use the ratio for the part of SOH CAH TOA that has two ticks.

Example 3.10

Calculate the size of the angle $\theta°$ in this triangle.

Give your answer to 1 decimal place (1 d.p.).

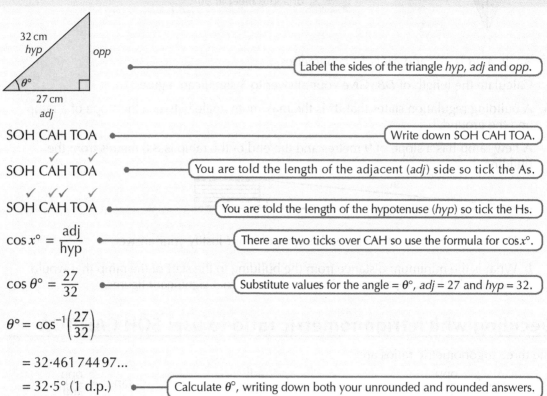

Label the sides of the triangle *hyp, adj* and *opp*.

SOH CAH TOA   Write down SOH CAH TOA.

SOH CAH TOA   You are told the length of the adjacent (*adj*) side so tick the As.

SOH CAH TOA   You are told the length of the hypotenuse (*hyp*) so tick the Hs.

$$\cos x° = \frac{adj}{hyp}$$ There are two ticks over CAH so use the formula for $\cos x°$.

$$\cos \theta° = \frac{27}{32}$$ Substitute values for the angle = $\theta°$, *adj* = 27 and *hyp* = 32.

$$\theta° = \cos^{-1}\left(\frac{27}{32}\right)$$

$$= 32{\cdot}461\,744\,97\ldots$$

$$= 32{\cdot}5° \text{ (1 d.p.)}$$ Calculate $\theta°$, writing down both your unrounded and rounded answers.

Example 3.11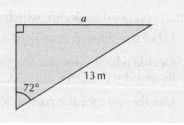

Calculate the length of the side marked *a* in the triangle.

Give your answer to 3 significant figures (3 s.f.).

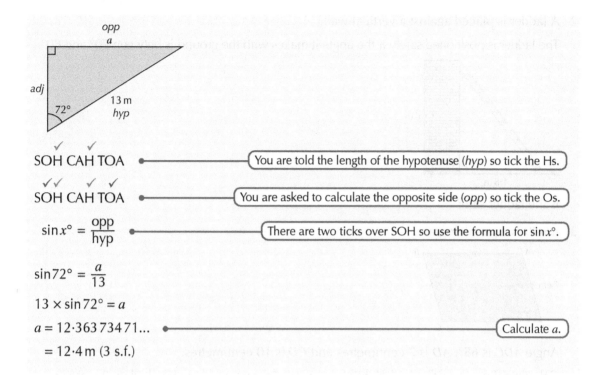

SOH CAH TOA ●──────── You are told the length of the hypotenuse (*hyp*) so tick the Hs.

SOH CAH TOA ●──────── You are asked to calculate the opposite side (*opp*) so tick the Os.

$\sin x° = \dfrac{\text{opp}}{\text{hyp}}$ ●──────── There are two ticks over SOH so use the formula for $\sin x°$.

$\sin 72° = \dfrac{a}{13}$

$13 \times \sin 72° = a$

$a = 12 \cdot 363\,734\,71...$ ●──────────────────────── Calculate *a*.

$= 12 \cdot 4\,\text{m}$ (3 s.f.)

N4 **Exercise 3E** 🖳

For Questions 1–5, give your answers to 3 significant figures (3 s.f.).

★ 1 Calculate the angle or side marked with a letter in the following triangles.

a

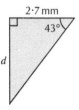

b

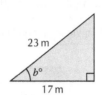

c

d

e

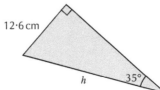

f

g

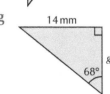

h

2 A ladder is placed against a vertical wall.

The ladder is positioned safely if the angle it makes with the ground is between 70° and 80°.

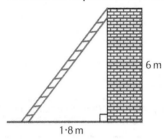

6 m

1·8 m

Is the ladder safe?

3 *ABCD* is a parallelogram.

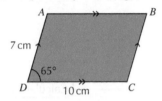

Angle *ADC* is 65°, *AD* is 7 centimetres and *CD* is 10 centimetres.

Calculate the perpendicular height of the parallelogram and hence calculate the area of the parallelogram.

> **Hint** The formula for calculating the area of a parallelogram is:
>
> area = base × perpendicular height

4 Calculate the area of an equilateral triangle with length of side 10 centimetres.

> **Hint** Draw the triangle. What size is each angle in an equilateral triangle?

5 a Calculate the size of angle *QPS*.

> **Hint** You may need to do more than one calculation in each of these questions.

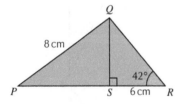

b Calculate the length *DC*.

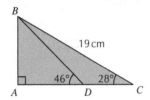

Chapter 3 review 🖩

For Questions 1–5, give all your answers to 1 decimal place (1 d.p.).

1 Calculate the angle or side marked with a letter in each triangle.

a
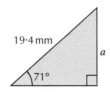

b
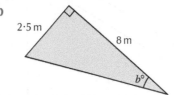

c
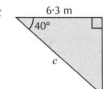

d
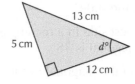

2 A garden security light is attached to a vertical wall 2·4 metres above the ground.
The light shines on a spot in the garden that is 3·3 metres away from the wall.

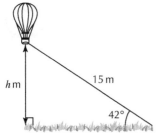

Calculate the size of angle $x°$.

3 At Strathaven Balloon Festival, a hot air balloon is tethered by a rope that is 15 metres long.

The angle of elevation (the angle between the ground and the rope) is 42°.

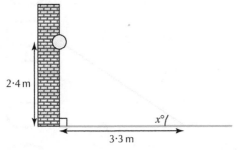

Calculate the height, h, of the balloon above the ground.

4 The highest point of a young child's slide is 1·1 metres above the ground.

   The angle between the slide and the ground is 38°.

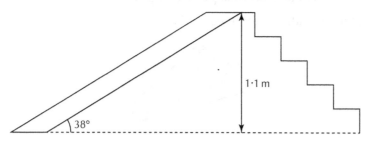

   What is the length of the slide?

5 The length of the longest side in a rectangle is 14 centimetres and the angle between the longest side and its diagonal is 38°.

   Calculate the length of the diagonal of the rectangle.

- I can use the tangent ratio to calculate the missing side in a right-angled triangle. ★ Exercise 3A Q1, Q2, Q5

- I can use the tangent ratio to calculate an angle in a right-angled triangle. ★ Exercise 3B Q1, Q4

- I can use the sine ratio to calculate the missing side in a right-angled triangle. ★ Exercise 3C Q1, Q3

- I can use the sine ratio to calculate an angle in a right-angled triangle. ★ Exercise 3C Q2, Q5

- I can use the cosine ratio to calculate the missing side in a right-angled triangle. ★ Exercise 3D Q1, Q4

- I can use the cosine ratio to calculate an angle in a right-angled triangle. ★ Exercise 3D Q2, Q5

- I can identify the different sides of a right-angled triangle and use this to choose the correct trig ratio. ★ Exercise 3E Q1

Geometry 1

This chapter will show you how to:

- use Pythagoras' theorem to find the length of the hypotenuse or one of the shorter sides given the other two sides in a right-angled triangle
- use the converse of Pythagoras' theorem to decide whether or not a triangle is right angled
- use Pythagoras' theorem in complex 2D situations
- calculate the size of an angle using angle properties of triangles and common quadrilaterals
- calculate the size of an angle using angle properties of circles
- solve problems using Pythagoras' theorem and the symmetry properties of circles.

You should already know:

- how to square and find the square root of a number
- how to round an answer to a required degree of accuracy
- how to plot points on a coordinate grid
- that the longest side of a right-angled triangle is called the hypotenuse and is opposite the right angle.

N4 Using Pythagoras' theorem to calculate the length of the hypotenuse in a right-angled triangle

Pythagoras' theorem is used to find the length of an unknown side in a right-angled triangle. The **longest side** in a right-angled triangle is called the **hypotenuse** and the hypotenuse is the side opposite the right angle.

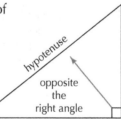

Important

Pythagoras' theorem states that:

If a triangle is right angled, then the area of the square on the hypotenuse is equal to the sum of the area of the squares on the other two sides.

$$c^2 = a^2 + b^2$$

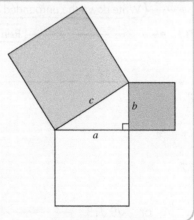

N4

Example 4.1 🖩

Find the length of the hypotenuse in each triangle. Give your answers as a surd or to 2 decimal places if required.

Hint
If you have completed Chapter 5, you will have worked with surds and can use them in these questions. Using surds is N5 work. Otherwise give your answer to 2 decimal places.

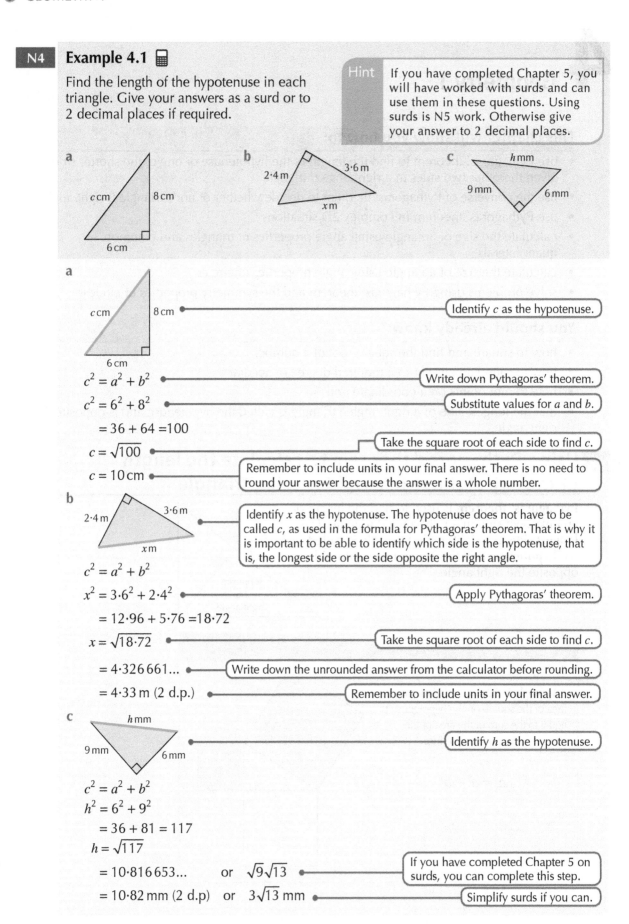

a

Identify c as the hypotenuse.

$c^2 = a^2 + b^2$ — Write down Pythagoras' theorem.

$c^2 = 6^2 + 8^2$ — Substitute values for a and b.

$= 36 + 64 = 100$

$c = \sqrt{100}$ — Take the square root of each side to find c.

$c = 10\,\text{cm}$ — Remember to include units in your final answer. There is no need to round your answer because the answer is a whole number.

b

Identify x as the hypotenuse. The hypotenuse does not have to be called c, as used in the formula for Pythagoras' theorem. That is why it is important to be able to identify which side is the hypotenuse, that is, the longest side or the side opposite the right angle.

$c^2 = a^2 + b^2$

$x^2 = 3\cdot6^2 + 2\cdot4^2$ — Apply Pythagoras' theorem.

$= 12\cdot96 + 5\cdot76 = 18\cdot72$

$x = \sqrt{18\cdot72}$ — Take the square root of each side to find c.

$= 4\cdot326\,661...$ — Write down the unrounded answer from the calculator before rounding.

$= 4\cdot33\,\text{m}$ (2 d.p.) — Remember to include units in your final answer.

c

Identify h as the hypotenuse.

$c^2 = a^2 + b^2$

$h^2 = 6^2 + 9^2$

$= 36 + 81 = 117$

$h = \sqrt{117}$

$= 10\cdot816\,653...$ or $\sqrt{9}\sqrt{13}$ — If you have completed Chapter 5 on surds, you can complete this step.

$= 10\cdot82\,\text{mm}$ (2 d.p)   or $3\sqrt{13}\,\text{mm}$ — Simplify surds if you can.

N4 **Exercise 4A** 🖩

For Questions 1–8, give your answers to 2 decimal places or as simplified surds.

★ 1 Calculate the length of the hypotenuse in the following triangles.

a
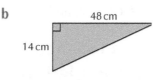

b
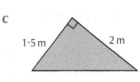

c

d
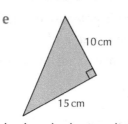

e
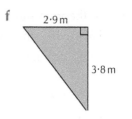

f

2 Which of the following rectangles has the longer diagonal?

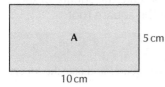
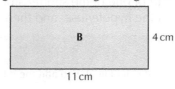

3 Calculate the length of a diagonal of a square that has sides 8 cm.

4 A ladder is placed against a vertical wall. The bottom of the ladder is at a horizontal distance of 0·6 m from the wall. The ladder reaches a height of 2·3 m up the wall.

Hint | Sketch a diagram.

Calculate the length of the ladder.

5 Daniel runs 6 km north then 5 km west. He then runs back directly to his starting point.

What distance has Daniel run altogether?

6 *PQRS* is a rhombus whose diagonals are 3·4 m and 2·6 m long.

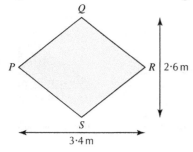

Calculate the perimeter of the rhombus.

7 Morgan's garden is shaped like a trapezium. He wants to surround his garden with a fence.

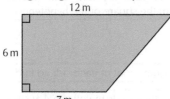

Calculate how many metres of fencing Morgan will need.

8 By plotting these points on coordinate axes, calculate the distance between each pair of points.

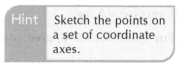

Hint Sketch the points on a set of coordinate axes.

a $A(3, 2)$ and $B(8, 9)$

b $C(2, 4)$ and $D(10, 8)$

c $E(-4, 8)$ and $F(10, -4)$

d $G(3, -2)$ and $H(-5, -7)$

N4 Using Pythagoras' theorem to calculate the length of one of the shorter sides in a right-angled triangle

Pythagoras' theorem states that if c is the hypotenuse of a right-angled triangle and a and b are the other two sides, then $c^2 = a^2 + b^2$

By rearranging Pythagoras' theorem to make a or b the subject of the formula, you get:

$$a^2 = c^2 - b^2 \qquad \text{or} \qquad b^2 = c^2 - a^2$$
$$a = \sqrt{c^2 - b^2} \qquad\qquad b = \sqrt{c^2 - a^2}$$

So if you want to find out the length of one of the shorter sides, you take the square of the other side away from the square of the hypotenuse, and then take the square root.

> **Important**
>
> - c is the hypotenuse or longest side; to calculate it, you **add** the other two squares and take the square root
> - a and b are shorter than c; to calculate either of them, **subtract** the square of a or b from c^2 and take the square root of the answer obtained to calculate the length of a or b.
>
>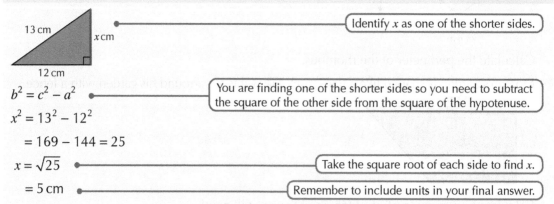

N4 Example 4.2

Calculate the length of the side marked x in the right-angled triangle.

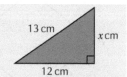

Identify x as one of the shorter sides.

$b^2 = c^2 - a^2$

You are finding one of the shorter sides so you need to subtract the square of the other side from the square of the hypotenuse.

$x^2 = 13^2 - 12^2$

$\quad = 169 - 144 = 25$

$x = \sqrt{25}$

Take the square root of each side to find x.

$\quad = 5\,cm$

Remember to include units in your final answer.

N4 **Example 4.3** 🖩

A ladder is 2·7 metres long and is placed against a vertical wall. The foot of the ladder is at a horizontal distance of 1 metre away from the bottom of the wall.

How far up the wall does the ladder reach? Give your answer to 1 decimal place.

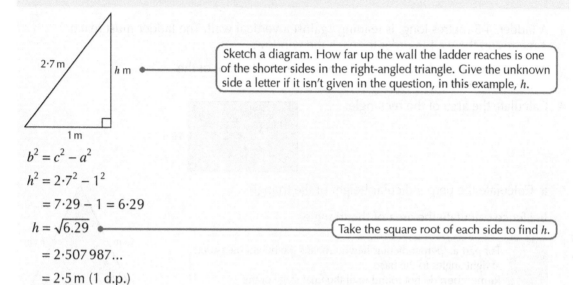

Sketch a diagram. How far up the wall the ladder reaches is one of the shorter sides in the right-angled triangle. Give the unknown side a letter if it isn't given in the question, in this example, h.

$b^2 = c^2 - a^2$

$h^2 = 2 \cdot 7^2 - 1^2$

$= 7 \cdot 29 - 1 = 6 \cdot 29$

$h = \sqrt{6.29}$ ⟵ Take the square root of each side to find h.

$= 2 \cdot 507\,987\ldots$

$= 2 \cdot 5\,\text{m (1 d.p.)}$

The ladder reaches 2·5 metres up the wall.

N4 **Exercise 4B** 🖩

For Questions 1–7, give your answers to 1 decimal place or as simplified surds.

★ 1 Calculate the length of the side marked x in each triangle.

a
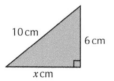
10 cm
6 cm
x cm

b
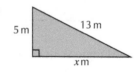
5 m
13 m
x m

c
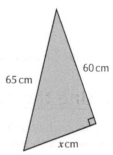
65 cm
60 cm
x cm

d
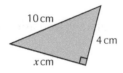
10 cm
4 cm
x cm

e
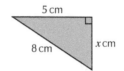
5 cm
8 cm
x cm

f
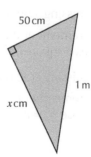
50 cm
1 m
x cm

2 Calculate the length marked *AB* in the diagram.

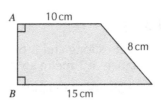

3 A ladder, 4·5 metres long, is leaning against a vertical wall. The ladder must reach 3 metres up the wall in order to reach the window.

At what horizontal distance from the wall should the foot of the ladder be placed?

4 Calculate the area of the rectangle.

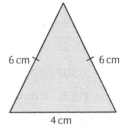

5 a Calculate the perpendicular height of the triangle.

b Hence calculate the area of the triangle.

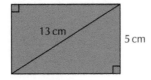

> **Hint** For part **a**, 'perpendicular height' means the height measured at right angles to the base.
> Remember: do not round until the final stage of the calculation.

6 Calculate the area of the triangle.

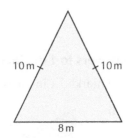

7 Calculate the area of an equilateral triangle with sides 8 cm.

N4 Choosing the correct method

When calculating the length of an unknown side in a right-angled triangle, whether it is the hypotenuse or a shorter side, you need to decide which method to use.

At the start of each question ask yourself:

- Is the missing side the hypotenuse?

  If it is, remember that the hypotenuse is the longest side so, to calculate it, you **add** the other two squares, that is, use $c^2 = a^2 + b^2$

- Is the missing side one of the shorter sides?

  If it is, remember that these sides are shorter than the hypotenuse so you **subtract** from c^2, that is, use $b^2 = c^2 - a^2$ or $a^2 = c^2 - b^2$

N4 **Exercise 4C**

★ **1** Find the length of the missing side in each triangle. Give your answers to 1 decimal place.

a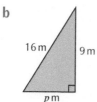

x cm

3 cm

5 cm

b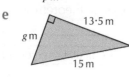

16 m

9 m

p m

c

7·1 cm

8·2 cm

d cm

d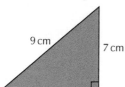

8 cm

t cm

5 cm

e

g m

13·5 m

15 m

f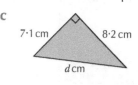

23 mm

20 mm

r mm

2 A ladder is 6 metres long. The bottom of the ladder is 2·5 metres horizontally from the foot of a vertical wall and the top leans against the wall.

How high is the top of the ladder above the ground? Give your answer to 1 decimal place.

3 Janice walks 300 m due south and then 400 m due west.

How far is Janice from her starting position?

4 Calculate the perimeter of the triangle. Give your answer to 2 decimal places.

9 cm

7 cm

N5 ## The converse of Pythagoras' theorem

Pythagoras' theorem states that:

> If a triangle is right angled, then the area of the square on the hypotenuse is equal to the sum of the area of the squares on the other two sides.

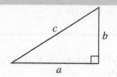

c

b

a

$$c^2 = a^2 + b^2$$

That is, the rule $c^2 = a^2 + b^2$ can only be used if you have a right-angled triangle.

You can use the **converse of Pythagoras' theorem** to show that you have a right-angled triangle.

Important

The **converse of Pythagoras' theorem** states that:

If $c^2 = a^2 + b^2$, then the triangle is right angled.

c

b

a

Example 4.4

Is the triangle right angled?

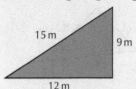

If the triangle is right angled, then Pythagoras' theorem will apply.

$15^2 = 225$ ——————————————— (Calculate the square of the longest side.)

$12^2 + 9^2 = 144 + 81 = 225$ ——— (Calculate the sum of the squares of the other two sides.)

$15^2 = 12^2 + 9^2$ ——————————— (Check if the above sums are equal.)

Hence by the converse of Pythagoras' theorem, the triangle is right angled. ——— (Make a statement that links the above sums and state your conclusion, mentioning the converse of Pythagoras' theorem **and** whether the triangle is right angled or not.)

Example 4.5

Katie is going to plant a rectangular lawn and she has marked out what she thinks is a rectangle in her garden. The sides of the area she has marked out for lawn measure 5·9 m and 4·5 m and the diagonal measures 7·5 m.

Will the lawn be rectangular?

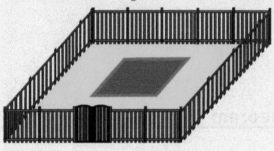

(Sketch and label the triangle.)

If the lawn is rectangular, it can be divided into two right angled triangles, and then Pythagoras' theorem will apply.

$7·5^2 = 56·25$

$4·5^2 + 5·9^2 = 20·25 + 34·81 = 55·06$

$7·5^2 \neq 4·5^2 + 5·9^2$ ——————— (Check if the above sums are equal. The symbol ≠ means 'is not equal to'.)

Hence by the converse of Pythagoras' theorem, the triangle is not right angled. ——— (State your conclusion in the context of the question.)

Katie's lawn will not be rectangular.

Exercise 4D

★ 1 Determine whether the following triangles are right angled or not.

> Hint   Remember to show all of your working and state your conclusion clearly.

a
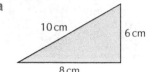
10 cm
6 cm
8 cm

b
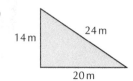
24 m
14 m
20 m

c
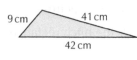
9 cm
41 cm
42 cm

d
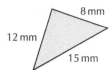
8 mm
12 mm
15 mm

e
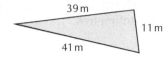
39 m
11 m
41 m

f
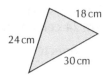
18 cm
24 cm
30 cm

★ 2 Are the following shapes rectangular?

a
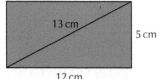
13 cm
5 cm
12 cm

b
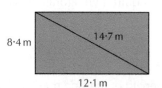
8·4 m
14·7 m
12·1 m

3 James is making a picture frame. The two sides of the frame are 22·5 cm and 12 cm and the diagonal is 25·5 cm.

Is the frame rectangular?

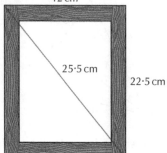
12 cm
25·5 cm
22·5 cm

4 To pass quality control inspection, the sides of a shed must be rectangular.

The two sides of the shed both measure 2·4 metres high by 3·2 metres long.
The diagonal of the side measures 4 metres.

Will the shed pass quality control?

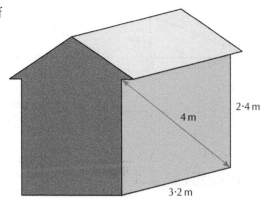
4 m
2·4 m
3·2 m

5 An isosceles triangle has two sides of 7 cm and a third side of 10 cm.

Is the triangle a right-angled isosceles triangle?

6   When marking out a football pitch, the
    groundsperson measures the diagonal of the
    pitch to check if the corners are right angles.

    The length of the pitch is 100 metres and the
    width is 67 metres. The diagonal measures
    120·45 metres.

    Is the pitch rectangular?

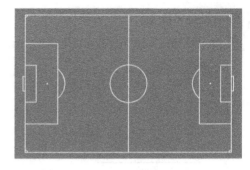

7   Marco draws a shape which he claims is a rhombus of side 12·5 centimetres.
    Charlotte says Marco is wrong. She measures the diagonals of the shape and finds
    they are 24 centimetres and 8 centimetres.

    Who is correct?

N5 Applying Pythagoras' theorem in complex 2D situations

Some problems require Pythagoras' theorem to be applied more than once, as illustrated by
Example 4.6.

N5 Example 4.6 🔢

Calculate the length of the side marked *AB*. Give your answer as a surd or to 1 decimal place.

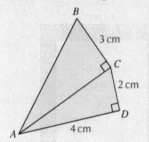

| Hint | If you have completed Chapter 5, you will have worked with surds and can use them in these questions. |

The shape is made of two right-angled triangles: triangle *ACD* and triangle *ABC*.

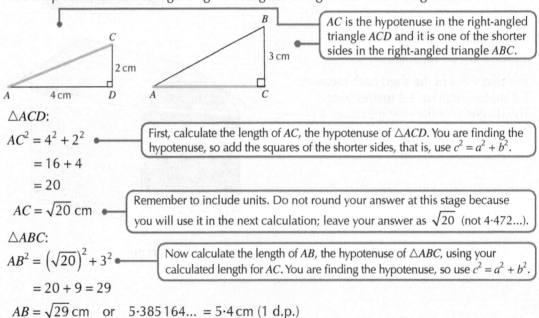

AC is the hypotenuse in the right-angled
triangle *ACD* and it is one of the shorter
sides in the right-angled triangle *ABC*.

$\triangle ACD$:

$AC^2 = 4^2 + 2^2$

First, calculate the length of *AC*, the hypotenuse of $\triangle ACD$. You are finding the
hypotenuse, so add the squares of the shorter sides, that is, use $c^2 = a^2 + b^2$.

$= 16 + 4$

$= 20$

$AC = \sqrt{20}$ cm

Remember to include units. Do not round your answer at this stage because
you will use it in the next calculation; leave your answer as $\sqrt{20}$ (not 4·472...).

$\triangle ABC$:

$AB^2 = \left(\sqrt{20}\right)^2 + 3^2$

Now calculate the length of *AB*, the hypotenuse of $\triangle ABC$, using your
calculated length for *AC*. You are finding the hypotenuse, so use $c^2 = a^2 + b^2$.

$= 20 + 9 = 29$

$AB = \sqrt{29}$ cm   or   5·385 164... = 5·4 cm (1 d.p.)

Exercise 4E 🖩

★ 1 Calculate the length of *x* and **then** *y* in each diagram. Give your answers to 1 decimal place.

> Hint It may help to sketch each diagram as two separate triangles and mark on the length of *x* once you have calculated it, as in Example 4.6.

a
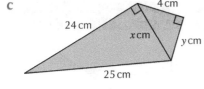

b
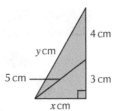

> Hint In part **b**, *y* is one of the shorter sides in the right-angled triangle so use $b^2 = c^2 - a^2$.

c
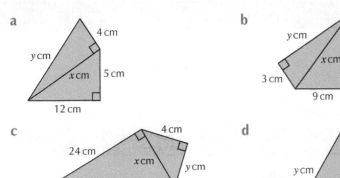

d
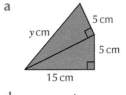

★ 2 Calculate the length of *y* in each diagram. Give your answers to 2 decimal places.

a
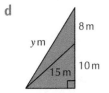

b
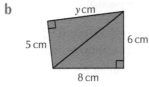

c
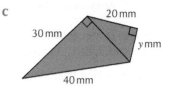

d
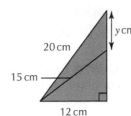

e

3 A vertical telephone mast is supported on one side by two cables.

Find the height of the mast and the length of the longer cable. Give your answer to 2 decimal places.

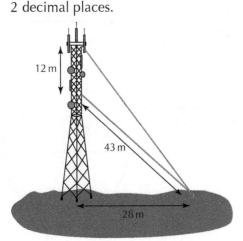

N4 Using parallel lines, symmetry and circle properties to calculate angles

Straight lines, parallel and intersecting lines and angles around a point have a range of angle properties as shown in the table.

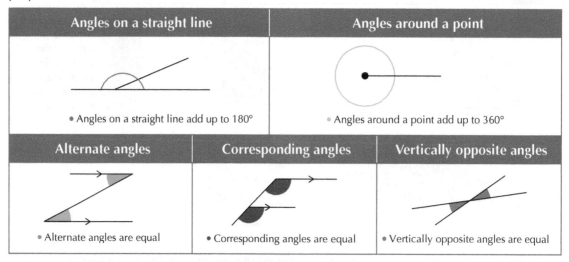

Angles on a straight line	Angles around a point
• Angles on a straight line add up to 180°	• Angles around a point add up to 360°

Alternate angles	Corresponding angles	Vertically opposite angles
• Alternate angles are equal	• Corresponding angles are equal	• Vertically opposite angles are equal

Hint   Arrows on lines show that they are parallel.

You can use combinations of these properties to find angles in intersecting and parallel lines, and in polygons. As well as the properties above, you can use these facts:

- angles in a triangle add to 180°
- isosceles triangles have two equal angles
- angles in a quadrilateral (4-sided shape) add to 360°.

A triangle formed by a chord and two radii
A line joining any two points on the circumference of a circle is a **chord**.

In the diagram, *AB* is a chord of the circle and *O* is the centre.

AOB is a triangle formed by the chord *AB* and the two radii *AO* and *BO*.

The two radii must be equal so the triangle *AOB* is isosceles.

Any triangle formed by a chord and two radii is isosceles.

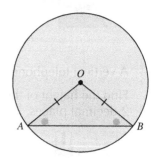

Angle in a semi-circle
AB is a diameter of a circle with centre *O*, splitting the circle into two semi-circles. *C* is a point on the circumference. ∠*ACB* is called the 'angle in a semi-circle'.

The angle in a semi-circle is a right angle.

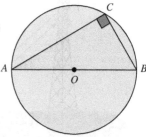

Angle between a tangent and the radius drawn to the point of contact

A line which touches a circle at exactly one point is a **tangent** to the circle.

If O is the centre of a circle and AC is a tangent to the circle touching at B, then the radius OB is perpendicular to the tangent AC.

The angle between a tangent and the radius drawn to the point of contact is a right angle.

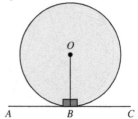

Important

- Any triangle formed by a chord and two radii is isosceles.
- The angle in a semi-circle is a right angle.
- The angle between a tangent and the radius drawn to the point of contact is a right angle.

N4

Example 4.7

a Work out the size of each angle marked with a letter.

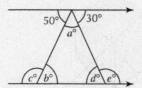

b O is the centre of the circle.

Calculate angles:

i $\angle ABC$

ii $\angle ACB$

iii $\angle AOB$.

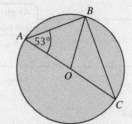

c O is the centre of a circle and ABC is a tangent to the circle.

Calculate:

i $\angle OBD$

ii $\angle DOB$.

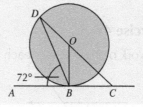

a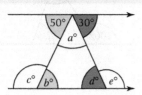

$50° + a° + 30° = 180°$ — Angles on a straight line add up to 180°.

$a° = 180° - 50° - 30° = 100°$

$b° = 50°$ — Angles $b°$ and 50° are alternate angles (highlighted blue).

$c° + b° = 180$ — Angles on a straight line add up to 180°.

$c° = 180° - 50° = 130°$

$d° = 30°$ — Angles $d°$ and 30° are alternate angles (highlighted red).

$d° + e° = 180°$ — Angles on a straight line add up to 180°.

$e° = 180° - 30° = 150°$

b

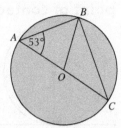

i ∠ABC = 90° •————————— Angle in a semi-circle is a right angle.

ii ∠ACB + ∠ABC + ∠BAC = 180° •————— Angles in a triangle add to 180°.

∠ACB = 180° − 90° − 53° = 37°

iii ∠ABO = ∠BAO = 53° •————— △AOB is an isosceles triangle because OA and OB are both radii.

∠OAB + ∠ABO + ∠AOB = 180° •————— Angles in a triangle add to 180°.

∠AOB = 180° − 53° − 53° = 74°

c

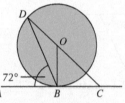

i 72° + ∠OBD = 90° •————— Angle between a tangent and the radius at the point of contact is a right angle.

∠OBD = 90° − 72° = 18°

ii ∠BDO = ∠OBD = 18° •————— △DOB is an isosceles triangle

∠BDO + ∠OBD + ∠DOB = 180°

∠DOB = 180° − 18° − 18° = 144°

N4 **Exercise 4F**

★ **1** Work out the size of each angle marked with a letter.

a

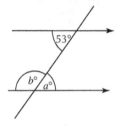

b

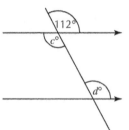

c

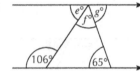

d

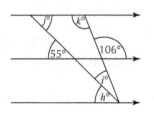

e

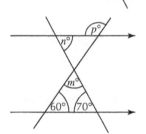

f

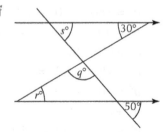

★ 2 Work out the size of each angle marked with a letter.

a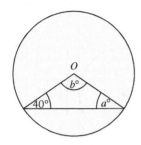

b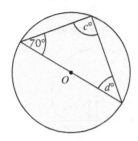

c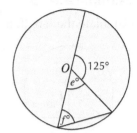

★ 3 Work out the size of each angle marked with a letter.

a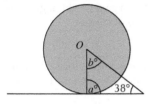

b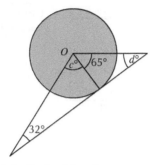

c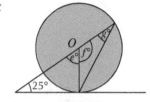

4 Work out the length of each side labelled with a letter.

a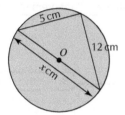

b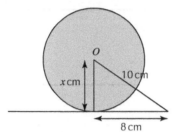

c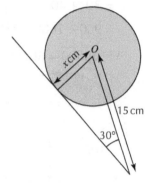

N5 ## Symmetry in the circle

The previous section showed that the triangle formed by two radii and a chord is isosceles.

Here are three more properties arising from the symmetry of shapes inscribed in (drawn in) circles:

- the perpendicular bisector of a chord passes through the centre of the circle

- a radius perpendicular to a chord bisects it

- a radius which bisects a chord is perpendicular to it.

A right-angled triangle is formed, so you can use Pythagoras' theorem to solve problems involving such shapes.

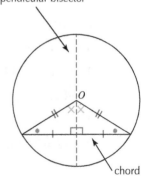

perpendicular bisector

O

chord

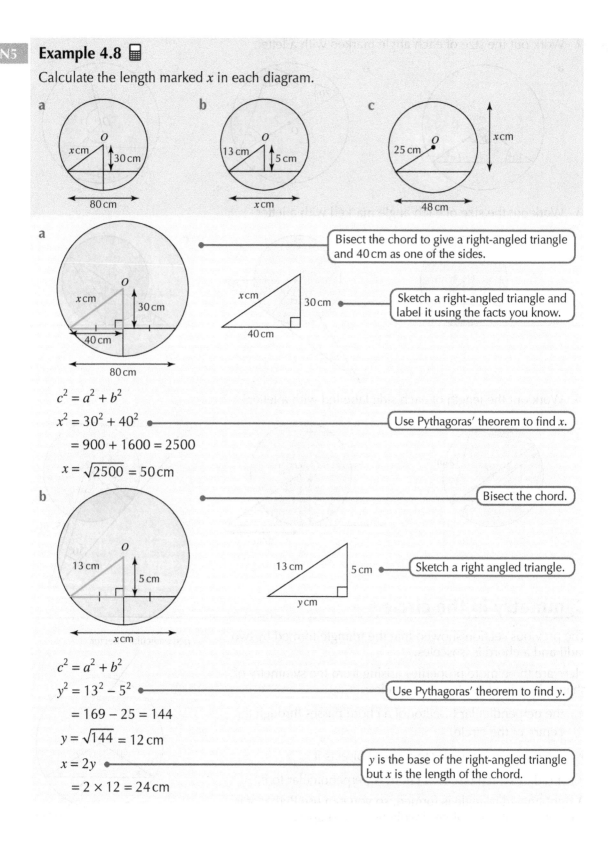

N5 **Example 4.8** 🖩

Calculate the length marked x in each diagram.

a

$x$cm
O
30 cm
80 cm

b

13 cm
O
5 cm
$x$cm

c

25 cm
O
$x$cm
48 cm

a

$x$cm
O
30 cm
40 cm
80 cm

> Bisect the chord to give a right-angled triangle and 40 cm as one of the sides.

$x$cm
30 cm
40 cm

> Sketch a right-angled triangle and label it using the facts you know.

$c^2 = a^2 + b^2$

$x^2 = 30^2 + 40^2$

> Use Pythagoras' theorem to find x.

$= 900 + 1600 = 2500$

$x = \sqrt{2500} = 50$ cm

b

13 cm
O
5 cm
$x$cm

> Bisect the chord.

13 cm
5 cm
$y$cm

> Sketch a right angled triangle.

$c^2 = a^2 + b^2$

$y^2 = 13^2 - 5^2$

> Use Pythagoras' theorem to find y.

$= 169 - 25 = 144$

$y = \sqrt{144} = 12$ cm

$x = 2y$

> y is the base of the right-angled triangle but x is the length of the chord.

$= 2 \times 12 = 24$ cm

c

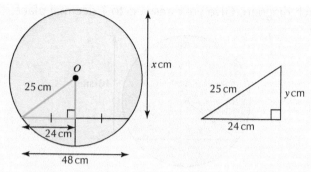

$$y^2 = 25^2 - 24^2$$ ●────────────────────(Use Pythagoras' theorem.)

$$= 625 - 576 = 49$$

$$y = \sqrt{49} = 7 \text{ cm}$$

$$x = y + 25$$ ●────────────────(To find x, add the length of the radius to y.)

$$= 32 \text{ cm}$$

N5 **Exercise 4G** 🖩

★ **1** Calculate the length marked x in each diagram. Give your answers to 1 decimal place.

a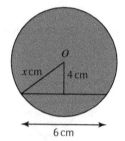

6 cm

b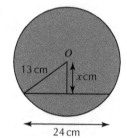

24 cm

c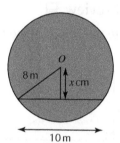

10 m

★ **2** Calculate the length marked x in each diagram. Give your answers to 1 decimal place.

a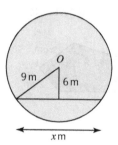

b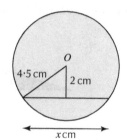

c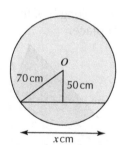

★ **3** Calculate the length marked x in each diagram. Give your answers to 1 decimal place.

a

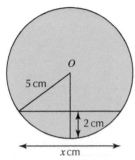

b

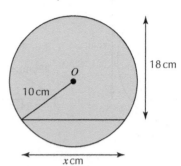

c

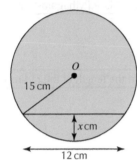

d

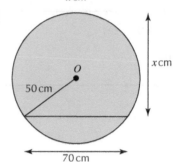

Chapter 4 review 🖩

1 Calculate the length of the missing side in each triangle. Give your answers to 2 decimal places.

a

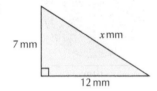

b

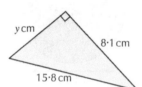

c

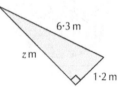

2 A triangle has sides 4·5 cm, 20 cm and 20·5 cm. Is the triangle right-angled?

3 Calculate the length of the side marked x in each diagram. Give your answers to 1 decimal place.

a

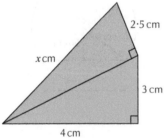

b

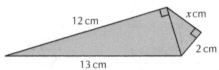

4 Work out the size of each angle marked with a letter.

a

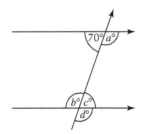

b

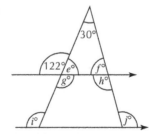

5 Work out the size of each angle marked with a letter.

a
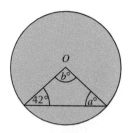

b
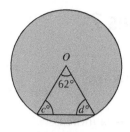

c
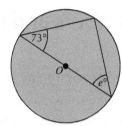

d
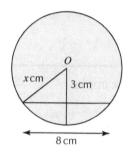

e
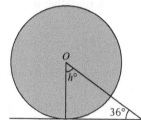

6 Calculate the length marked x in each diagram. Give your answers to 1 decimal place.

a
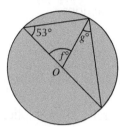

b
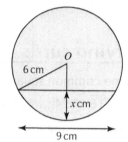

c
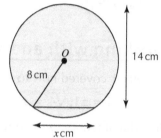

- I can use Pythagoras' theorem to find the length of the hypotenuse or one of the shorter sides given the other two sides in a right-angled triangle. ★ Exercise 4A Q1 ★ Exercise 4B Q1 ★ Exercise 4C Q1

- I can use the converse of Pythagoras' theorem to decide whether or not a triangle is right angled. ★ Exercise 4D Q1, Q2

- I can use Pythagoras' theorem in complex 2D situations. ★ Exercise 4E Q1, Q2

- I can calculate the size of an angle using angle properties of triangles and common quadrilaterals. ★ Exercise 4F Q1

- I can calculate the size of an angle using angle properties of circles. ★ Exercise 4F Q2, Q3

- I can solve problems using Pythagoras' theorem and the symmetry properties of circles. ★ Exercise 4G Q1–Q3

5 Numbers and indices 2

This chapter will show you how to:

- work with surds and simplify them
- simplify expressions using the rules of indices
- multiply and divide using positive and negative indices including fractions.

You should already know how to:

- carry out basic calculations involving integers (positive and negative whole numbers and zero) (see Chapter 1 pages 1–3)
- carry out numerical calculations involving fractions (addition, subtraction, multiplication and division) (see Chapter 1 pages 4–8)
- carry out numerical calculations with basic indices (or powers) and roots (see Chapter 1 pages 9–10).

N5 Working with and simplifying surds

Chapter 1 covered how to evaluate some common square roots, for example:

$$\sqrt{1} = \pm1, \ \sqrt{4} = \pm2, \ \sqrt{9} = \pm3, \ \sqrt{16} = \pm4, \ ... \ \sqrt{100} = \pm10, \ ...$$

But what about the square roots of the whole numbers in between, such as:

$$\sqrt{2}, \ \sqrt{3}, \ \sqrt{5}, \ \sqrt{6}, \ ... \ \sqrt{99}, \ ... \ \sqrt{101}, \ ... \text{ and so on?}$$

Using a calculator to evaluate these square roots gives:

$$\sqrt{2} = 1\cdot4142...$$

$$\sqrt{3} = 1\cdot7320...$$

$$\sqrt{5} = 2\cdot2360...$$

$$\sqrt{6} = 2\cdot4494...$$

$$\sqrt{99} = 9\cdot9498...$$

$$\sqrt{101} = 10\cdot0498...$$

and so on.

From this you can see that many square roots do not have whole-number answers, but instead are non-recurring decimal fractions.

The square roots of whole numbers which have non-recurring decimal answers are called **surds**. So, the surds are:

$$\sqrt{2}, \ \sqrt{3}, \ \sqrt{5}, \ \sqrt{6}, \ \sqrt{7}, \ \sqrt{8}, \ \sqrt{10}, \ \sqrt{11}, \ \sqrt{12}, \ \sqrt{13}, \ \sqrt{14}, \ \sqrt{15}, \ \sqrt{17}, ...$$

Some square roots are not included in this list because they are **not** surds, but instead are whole numbers:

$$\sqrt{1} = 1, \ \sqrt{4} = 2, \ \sqrt{9} = 3, \ \sqrt{16} = 4, \ ... \ \sqrt{100} = 10, \ ...$$

> **Hint** For all your work on surds, try to remember the following square roots:
>
> $$\sqrt{1} = 1 \qquad \sqrt{4} = 2 \qquad \sqrt{9} = 3 \qquad \sqrt{16} = 4 \qquad \sqrt{25} = 5$$
> $$\sqrt{36} = 6 \qquad \sqrt{49} = 7 \qquad \sqrt{64} = 8 \qquad \sqrt{81} = 9 \qquad \sqrt{100} = 10$$
> $$\sqrt{121} = 11 \qquad \sqrt{144} = 12 \qquad \sqrt{169} = 13$$
>
> It also useful to know (or be able to deduce) these common square roots:
>
> $$\sqrt{225} = 15 \qquad \sqrt{400} = 20 \qquad \sqrt{625} = 25 \qquad \sqrt{900} = 30 \qquad \sqrt{1600} = 40$$

N5 Adding and subtracting surds

You can add and subtract multiples of the **same** surd, for example:

$$5\sqrt{2} + \sqrt{2} = 6\sqrt{2} \qquad \text{and} \qquad 5\sqrt{2} - \sqrt{2} = 4\sqrt{2}$$

Consider the surds $\sqrt{7}$ and $\sqrt{3}$. Using a calculator, you can see that:

$$\sqrt{7} + \sqrt{3} = 4.3778...$$

Using a calculator, you can also see that:

$$\sqrt{7 + 3} = \sqrt{10} = 3.1622...$$

From this you can see that:

$$\sqrt{7} + \sqrt{3} \neq \sqrt{7 + 3}$$

> **Hint** The symbol $\neq$ means 'is not equal to'.

Similarly, using a calculator shows that:

$$\sqrt{7} - \sqrt{3} = 0.9137...$$

whereas:

$$\sqrt{7 - 3} = \sqrt{4} = 2$$

From this you can see that:

$$\sqrt{7} - \sqrt{3} \neq \sqrt{7 - 3}$$

So, there doesn't appear to be a rule for adding or subtracting **different** surds together, such as $\sqrt{7}$ and $\sqrt{3}$.

Rule 1: Multiplying surds

Using a calculator, you can see that:

$$\sqrt{7} \times \sqrt{3} = 4.5825...$$

and also:

$$\sqrt{7 \times 3} = \sqrt{21} = 4.5825...$$

From this you can see that:

$$\sqrt{7} \times \sqrt{3} = \sqrt{7 \times 3} = \sqrt{21}$$

Repeating the comparison with another pair of surds, for example, $\sqrt{5}$ and $\sqrt{12}$, gives:

$$\sqrt{5} \times \sqrt{12} = 7.7459... \qquad \text{and} \qquad \sqrt{5 \times 12} = \sqrt{60} = 7.7459...$$

That is:

$$\sqrt{5} \times \sqrt{12} = \sqrt{5 \times 12} = \sqrt{60}$$

These two examples illustrate the first rule of surds:

> **Important**
>
> **Rule 1:** If x and y are whole numbers then:
>
> $$\sqrt{x}\sqrt{y} = \sqrt{xy}$$

N5 **Example 5.1**

a Express as single surds:

   i $\sqrt{2} \times \sqrt{10}$

   ii $\sqrt{3} \times \sqrt{5} \times \sqrt{10}$

b Express as a product of two or more surds:

   i $\sqrt{2 \times 7}$

> **Hint** Remember that in maths, the word **product** means 'multiply' or 'times'.

   ii $\sqrt{3 \times 13 \times 19}$

a i $\sqrt{2} \times \sqrt{10} = \sqrt{2 \times 10} = \sqrt{20}$

   ii $\sqrt{3} \times \sqrt{5} \times \sqrt{10} = \sqrt{3 \times 5 \times 10} = \sqrt{150}$

b i $\sqrt{2 \times 7} = \sqrt{2} \times \sqrt{7} = \sqrt{2}\sqrt{7}$

> In such expressions, you omit the × sign in the same way as you omit it in algebra, e.g. $3 \times y = 3y$

   ii $\sqrt{3 \times 13 \times 19} = \sqrt{3} \times \sqrt{13} \times \sqrt{19}$

> The rule can be applied to more than two surds.

        $= \sqrt{3}\sqrt{13}\sqrt{19}$

N5 **Exercise 5A**

★ 1 Express as single surds:

   a $\sqrt{3} \times \sqrt{5}$ b $\sqrt{2} \times \sqrt{13}$ c $\sqrt{6} \times \sqrt{7}$

   d $\sqrt{30} \times \sqrt{40}$ e $\sqrt{2} \times \sqrt{7} \times \sqrt{10}$ f $\sqrt{2} \times \sqrt{6} \times \sqrt{20}$

2 Express as a product of two or more surds:

   a $\sqrt{5 \times 7}$ b $\sqrt{11 \times 13}$

   c $\sqrt{2 \times 29 \times 23}$ d $\sqrt{3 \times 17 \times 101}$

N5 **Multiplying surds further**

In many instances you will need to simplify a single surd by expressing it as a product of two or more simpler, smaller surds or whole numbers. In order to do this, it is useful to think of Rule 1 'in reverse', that is:

$$\sqrt{xy} = \sqrt{x}\sqrt{y}$$

Then, if $\sqrt{x}$ or $\sqrt{y}$ is a whole number, you might be able to simplify further. Always look for the **greatest square number** that is a factor of the given number within the square root.

N5 **Example 5.2**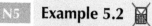

a Simplify:

   i $\sqrt{12}$ **ii** $\sqrt{125}$ **iii** $\sqrt{700}$

b Simplify:

   i $\sqrt{32}$ **ii** $\sqrt{200}$

c Hence simplify:

   i $\sqrt{32} + \sqrt{200} - 5\sqrt{2}$ **ii** $8\sqrt{5} - \sqrt{80} - \sqrt{45}$ **iii** $\sqrt{700} + \sqrt{28}$

> **Hint**   **Hence** means use the answers you have just worked out in order to answer the new question.

a **i** $\sqrt{12} = \sqrt{4 \times 3} = \sqrt{4} \times \sqrt{3}$

> The square numbers which are factors of 12 are 1 and 4, but 4 is greater.

        $= 2 \times \sqrt{3} = 2\sqrt{3}$

 ii $\sqrt{125} = \sqrt{25 \times 5} = \sqrt{25} \times \sqrt{5}$

> The square numbers which are factors of 125 are 1 and 25, but 25 is greater.

        $= 5 \times \sqrt{5} = 5\sqrt{5}$

 iii $\sqrt{700} = \sqrt{100 \times 7} = \sqrt{100} \times \sqrt{7}$

> The square numbers which are factors of 700 are 1, 25 and 100, but 100 is the greatest.

        $= 10 \times \sqrt{7} = 10\sqrt{7}$

b **i** $\sqrt{32} = \sqrt{16 \times 2}$

> The square numbers which are factors of 32 are 1, 4 and 16 but the greatest is 16.

        $= \sqrt{16}\sqrt{2}$

        $= 4\sqrt{2}$

 ii $\sqrt{200} = \sqrt{100 \times 2}$

        $= \sqrt{100}\sqrt{2} = 10\sqrt{2}$

c **i** $\sqrt{32} + \sqrt{200} - 5\sqrt{2} = 4\sqrt{2} + 10\sqrt{2} - 5\sqrt{2}$

> You can add and subtract surds which are of the same type.

        $= 9\sqrt{2}$

 ii $8\sqrt{5} - \sqrt{80} - \sqrt{45} = 8\sqrt{5} - \sqrt{16 \times 5} - \sqrt{9 \times 5}$

        $= 8\sqrt{5} - \sqrt{16}\sqrt{5} - \sqrt{9}\sqrt{5}$

        $= 8\sqrt{5} - 4\sqrt{5} - 3\sqrt{5} = \sqrt{5}$

 iii $\sqrt{700} + \sqrt{28} = \sqrt{100 \times 7} + \sqrt{4 \times 7} = \sqrt{100}\sqrt{7} + \sqrt{4}\sqrt{7}$

        $= 10\sqrt{7} + 2\sqrt{7} = 12\sqrt{7}$

N5 **Exercise 5B**

★ 1 Simplify:

a $\sqrt{8}$ b $\sqrt{75}$ c $\sqrt{45}$ d $\sqrt{28}$

e $\sqrt{32}$ f $\sqrt{72}$ g $\sqrt{48}$ h $\sqrt{99}$

> **Hint** Remember to look for the largest square number that is a factor.

2 Simplify:

a $\sqrt{80}$ b $\sqrt{300}$ c $\sqrt{200}$ d $\sqrt{52}$

e $\sqrt{63}$ f $\sqrt{98}$ g $\sqrt{108}$ h $\sqrt{44}$

★ 3 Simplify:

a $10\sqrt{2} + \sqrt{8} - \sqrt{50}$ b $\sqrt{12} + \sqrt{27} - 4\sqrt{3}$

c $\sqrt{45} - \sqrt{5} - \sqrt{20}$ d $\sqrt{44} + \sqrt{99} + \sqrt{11}$

4 Simplify:

a $\sqrt{200} + \sqrt{8}$ b $\sqrt{45} - \sqrt{20}$ c $\sqrt{12} + \sqrt{48} - \sqrt{27}$

N5 **Rule 2: Dividing surds**

Consider the surds $\sqrt{80}$ and $\sqrt{5}$. Using a calculator, you can see that:

$$\frac{\sqrt{80}}{\sqrt{5}} = 4$$

You can also see that:

$$\sqrt{\frac{80}{5}} = \sqrt{16} = 4$$

From this you can see that:

$$\frac{\sqrt{80}}{\sqrt{5}} = \sqrt{\frac{80}{5}}$$

Repeating the comparison with another pair of surds, for example, $\sqrt{135}$ and $\sqrt{12}$, gives:

$$\frac{\sqrt{135}}{\sqrt{12}} = 3{\cdot}3541... \qquad \text{and} \qquad \sqrt{\frac{135}{12}} = 3{\cdot}3541...$$

That is:

$$\frac{\sqrt{135}}{\sqrt{12}} = \sqrt{\frac{135}{12}}$$

These two examples illustrate the second rule of surds:

> **Important**
>
> **Rule 2:** If x and y are whole numbers then:
> $$\frac{\sqrt{x}}{\sqrt{y}} = \sqrt{\frac{x}{y}}$$

N5 **Example 5.3** 🖩

a Express as single surds:

  i $\dfrac{\sqrt{21}}{\sqrt{5}}$ **ii** $\dfrac{\sqrt{101}}{\sqrt{2}}$ **iii** $\sqrt{3} \div \sqrt{50}$

b Express as a quotient of two surds:

> **Hint** In maths, the word **quotient** means division.

  i $\sqrt{\dfrac{47}{3}}$ **ii** $\sqrt{\dfrac{2}{11}}$ **iii** $\sqrt{40 \div 53}$

a i $\dfrac{\sqrt{21}}{\sqrt{5}} = \sqrt{\dfrac{21}{5}}$

  ii $\dfrac{\sqrt{101}}{\sqrt{2}} = \sqrt{\dfrac{101}{2}}$

  iii $\sqrt{3} \div \sqrt{50} = \dfrac{\sqrt{3}}{\sqrt{50}}$ ●———— Divisions can be written as fractions ('*a* over *b*').

$= \sqrt{\dfrac{3}{50}}$

b i $\sqrt{\dfrac{47}{3}} = \dfrac{\sqrt{47}}{\sqrt{3}}$

  ii $\sqrt{\dfrac{2}{11}} = \dfrac{\sqrt{2}}{\sqrt{11}}$

  iii $\sqrt{40 \div 53} = \sqrt{\dfrac{40}{53}}$

$= \dfrac{\sqrt{40}}{\sqrt{53}}$

N5 **Exercise 5C** 🖩

1 Express as single surds:

  a $\dfrac{\sqrt{56}}{\sqrt{3}}$ **b** $\dfrac{\sqrt{21}}{\sqrt{5}}$ **c** $\dfrac{\sqrt{2}}{\sqrt{7}}$ **d** $\dfrac{\sqrt{75}}{\sqrt{46}}$

  e $\sqrt{7} \div \sqrt{60}$ **f** $\sqrt{101} \div \sqrt{102}$ **g** $\sqrt{55} \div \sqrt{2}$

2 Express as a quotient of two surds:

  a $\sqrt{\dfrac{11}{2}}$ **b** $\sqrt{\dfrac{53}{21}}$ **c** $\sqrt{\dfrac{3}{34}}$ **d** $\sqrt{\dfrac{201}{202}}$

  e $\sqrt{20 \div 51}$ **f** $\sqrt{11 \div 31}$ **g** $\sqrt{80 \div 7}$

Evaluating surds without a calculator 🀫

Example 5.4 shows how to use rule 2 to evaluate expressions without using a calculator.

N5 **Example 5.4** 🀫

Evaluate:

a $\dfrac{\sqrt{60}}{\sqrt{15}}$ b $\sqrt{1400} \div \sqrt{14}$ c $\sqrt{\dfrac{144}{25}}$ d $\sqrt{16 \div 81}$

a $\dfrac{\sqrt{60}}{\sqrt{15}} = \sqrt{\dfrac{60}{15}}$

Use rule 2: $\dfrac{\sqrt{x}}{\sqrt{y}} = \sqrt{\dfrac{x}{y}}$

$= \sqrt{60 \div 15}$

$= \sqrt{4} = 2$

b $\sqrt{1400} \div \sqrt{14} = \sqrt{\dfrac{1400}{14}}$

$= \sqrt{1400 \div 14}$

$= \sqrt{100} = 10$

c $\sqrt{\dfrac{144}{25}} = \dfrac{\sqrt{144}}{\sqrt{25}}$

Use the reverse of rule 2: $\sqrt{\dfrac{x}{y}} = \dfrac{\sqrt{x}}{\sqrt{y}}$

$= \dfrac{12}{5}$

d $\sqrt{16 \div 81} = \sqrt{\dfrac{16}{81}} = \dfrac{\sqrt{16}}{\sqrt{81}} = \dfrac{4}{9}$

N5 **Exercise 5D** 🀫

1 Evaluate:

a $\dfrac{\sqrt{90}}{\sqrt{10}}$ b $\dfrac{\sqrt{48}}{\sqrt{12}}$ c $\dfrac{\sqrt{112}}{\sqrt{7}}$ d $\dfrac{\sqrt{75}}{\sqrt{3}}$

e $\sqrt{600} \div \sqrt{6}$ f $\sqrt{52} \div \sqrt{13}$ g $\sqrt{2800} \div \sqrt{7}$

★ 2 Evaluate:

a $\sqrt{\dfrac{81}{25}}$ b $\sqrt{\dfrac{4}{121}}$ c $\sqrt{\dfrac{100}{9}}$ d $\sqrt{\dfrac{1600}{169}}$

e $\sqrt{36 \div 49}$ f $\sqrt{225 \div 144}$ g $\sqrt{1 \div 900}$

N5 The rules of indices

Index is another word for 'power' (the plural of index is **indices**; see Chapter 1 page 9).

In the expression 5^2 (read as 'five squared'), the **base** is 5 and the **index** is 2:

base ⟶ 5^2 ⟵ index

Hint	Look back at Chapter 1, especially Example 1.6 and Exercise 1E (page 9) for more information on evaluating powers of numbers.

Previously when we have found the square root $\sqrt{}$ we have listed both the positive and negative solution. For this section on indices we will be finding the **principal square root**. That is, the positive root.

N5 Rule 1: Indices when multiplying

Consider $6^2 \times 6^3$:

$$6^2 \times 6^3 = (6 \times 6) \times (6 \times 6 \times 6)$$
$$= 6 \times 6 \times 6 \times 6 \times 6$$
$$= 6^5$$

Now consider $3^3 \times 3^5$:

$$3^3 \times 3^5 = (3 \times 3 \times 3) \times (3 \times 3 \times 3 \times 3 \times 3)$$
$$= 3 \times 3 \times 3 \times 3 \times 3 \times 3 \times 3 \times 3$$
$$= 3^8$$

Now consider an example where the base is a letter (or variable) rather than a number:

$$a^2 \times a^4 = (a \times a) \times (a \times a \times a \times a)$$
$$= a \times a \times a \times a \times a \times a$$
$$= a^6$$

So:

$$6^2 \times 6^3 = 6^{2+3} = 6^5 \qquad 3^3 \times 3^5 = 3^{3+5} = 3^8 \qquad a^2 \times a^4 = a^{2+4} = a^6$$

From this you can see that when you multiply powers of the same base together you **add the indices**.

This leads to the first rule of indices:

> **Important**
>
> **Rule 1:** $a^x \times a^y = a^{x+y}$

N5 Example 5.5 🖩

Express each of the following with a single index:

a $a^2 \times a^8$ **b** $x \times x^5$ **c** $4b^3 \times 10b^5$ **d** $2t^6 \times 3t^3$

a $a^2 \times a^8 = a^{2+8}$ ⟶ Add the powers.
$$= a^{10}$$

b $x \times x^5 = x^1 \times x^5$ ⟶ $x = x^1$
$$= x^{1+5} = x^6$$

c $4b^3 \times 10b^5 = 4 \times 10 \times b^3 \times b^5$ ⟶ Multiply the whole numbers together and multiply the variables together.
$$= 40b^{3+5} = 40b^8$$

d $2t^6 \times 3t^3 = 6t^{6+3} = 6t^9$

N5 **Exercise 5E**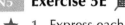

★ 1 Express each of the following with a single index:

a $x^3 \times x^5$ b $a^7 \times a^2$ c $k^6 \times k^7$ d $p^4 \times p^3$

e $t \times t^3$ f $w^{99} \times w$ g $5^6 \times 5^4$ h $7^{10} \times 7^5$

i $2^2 \times 2^2$ j $10^{11} \times 10$ k $3^u \times 3^v$ l $y^p \times y^q$

2 Express each of the following with a single index:

a $p^2 \times p^3 \times p^6$ b $a^{10} \times a^4 \times a$ c $t^4 \times t^4 \times t^4$ d $6 \times 6^3 \times 6^6$

e $2x^4 \times 5x^3$ f $3a^2 \times 3a$ g $4y^4 \times 3y^3$ h $3t^2 \times 4t^3 \times t^4$

i $4m \times 2m^3 \times 10m^4$ j $6p \times 2p^5 \times 10p^3$ k $5^p \times 5^q \times 5^r$ l $y^a \times y^b \times y^c$

Hint   You can apply rule 1 to a product of more than two powers, for example, $a^x \times a^y \times a^z = a^{x+y+z}$

N5 **Rule 2: Indices when dividing**

Consider $4^5 \div 4^2$:

$$4^5 \div 4^2 = \frac{4^5}{4^2} = \frac{\cancel{4} \times \cancel{4} \times 4 \times 4 \times 4}{\cancel{4} \times \cancel{4}}$$

$$= 4 \times 4 \times 4 = 4^3$$

Now consider $\dfrac{7^6}{7^4}$:

$$\frac{7^6}{7^4} = \frac{\cancel{7} \times \cancel{7} \times \cancel{7} \times \cancel{7} \times 7 \times 7}{\cancel{7} \times \cancel{7} \times \cancel{7} \times \cancel{7}}$$

$$= 7 \times 7 = 7^2$$

Now consider an example where the base is a variable rather than a number:

$$\frac{p^7}{p^3} = \frac{\cancel{p} \times \cancel{p} \times \cancel{p} \times p \times p \times p \times p}{\cancel{p} \times \cancel{p} \times \cancel{p}}$$

$$= p \times p \times p \times p = p^4$$

So:

$$\frac{4^5}{4^2} = 4^{5-2} = 4^3 \qquad \frac{7^6}{7^4} = 7^{6-4} = 7^2 \qquad \frac{p^7}{p^3} = p^{7-3} = p^4$$

From this you can see that when you divide powers of the same base together you **subtract the indices**.

This leads to the second rule of indices:

Important

Rule 2: $\dfrac{a^x}{a^y} = a^{x-y}$

N5 **Example 5.6**

a Express each of the following with a single index.

i $\dfrac{a^8}{a^3}$ ii $\dfrac{x^7}{x^6}$ iii $p^{10} \div p$

b Simplify and express each of the following with a single index.

i $20m^8 \div 5m^6$ ii $\dfrac{15t^8}{25t^7}$

a i $\dfrac{a^8}{a^3} = a^{8-3} = a^5$

ii $\dfrac{x^7}{x^6} = x^{7-6} = x$ ●————————————($x = x^1$)

iii $p^{10} \div p = \dfrac{p^{10}}{p^1}$

$= p^{10-1} = p^9$

b i $20m^8 \div 5m^6 = \dfrac{^4\cancel{20}m^8}{_1\cancel{5}m^6}$ ●————(Divide the whole numbers and the variables separately.)

$= 4m^{8-6} = 4m^2$

ii $\dfrac{15t^8}{25t^7} = \dfrac{^3\cancel{15}t^8}{_5\cancel{25}t^7}$

$= \dfrac{3t^{8-7}}{5} = \dfrac{3t}{5}$

N5 **Exercise 5F**

1 Express each of the following with a single index.

a $\dfrac{a^9}{a^2}$ b $\dfrac{p^{12}}{p^8}$ c $\dfrac{t^5}{t^4}$ d $\dfrac{w^6}{w}$

e $m^7 \div m^5$ f $k^8 \div k$ g $b^{11} \div b^{10}$ h $\dfrac{9^4}{9^2}$

i $7^7 \div 7^3$ j $\dfrac{8^{10}}{8^9}$ k $4^3 \div 4$

★ 2 Simplify and express each of the following with a single index.

a $\dfrac{30x^8}{6x^4}$ b $\dfrac{8y^5}{2y^2}$ c $12w^{10} \div 3w^3$

d $9a^6 \div 3a^5$ e $\dfrac{2t^6}{4t}$ f $\dfrac{3q^{12}}{30q^{10}}$

3 Simplify fully:

a $\dfrac{35b^7}{25b^5}$ b $\dfrac{16r^8}{24r^7}$ c $6g^7 \div 9g^2$ d $700a^6 \div 200a$

N5 **Rule 3: Raising a power to another power**

Consider $\left(x^2\right)^3$:

$$\left(x^2\right)^3 = x^2 \times x^2 \times x^2$$

$$= x^{2+2+2} = x^6$$

Also, for example:

$$\left(a^5\right)^4 = a^5 \times a^5 \times a^5 \times a^5$$

$$= a^{5+5+5+5} = a^{20}$$

So:

$$\left(x^2\right)^3 = x^{2 \times 3} = x^6 \qquad \text{and} \qquad \left(a^5\right)^4 = a^{5 \times 4} = a^{20}$$

In a similar way, it can be shown that, for example:

$$\left(y^{10}\right)^5 = y^{10 \times 5} = y^{50} \qquad \text{and} \qquad \left(6^3\right)^4 = 6^{3 \times 4} = 6^{12}$$

The examples illustrate the third rule of indices:

Important

Rule 3: $\left(a^x\right)^y = a^{xy}$

N5 **Example 5.7**

Express each of the following with a single index.

a $\left(a^4\right)^6$ b $\left(y^5\right)^2$ c $\left(8^2\right)^3 \times 8$ d $\left(p^3\right)^7 \times p^9$ e $\dfrac{\left(t^2\right)^6 \times \left(t^3\right)^3}{\left(t^5\right)^4}$

a $\left(a^4\right)^6 = a^{4 \times 6} = a^{24}$

b $\left(y^5\right)^2 = y^{5 \times 2} = y^{10}$

c $\left(8^2\right)^3 \times 8 = 8^6 \times 8^1 = 8^7$ •————————(Using rule 1, add powers when multiplying.)

d $\left(p^3\right)^7 \times p^9 = p^{21} \times p^9 = p^{30}$

e $\dfrac{\left(t^2\right)^6 \times \left(t^3\right)^3}{\left(t^5\right)^4} = \dfrac{t^{12} \times t^9}{t^{20}} = \dfrac{t^{21}}{t^{20}} = t$ •————(Using rule 2, subtract powers when dividing.)

N5 **Exercise 5G**

★ 1 Express each of the following with a single index.

a $\left(a^3\right)^8$ b $\left(x^6\right)^7$ c $\left(p^5\right)^5$ d $\left(6^5\right)^7$

e $\left(2^4\right)^4$ f $\left(x^4\right)^p$ g $\left(x^w\right)^3$ h $\left(x^a\right)^b$

2 Simplify and express each of the following with a single index.

a $\left(a^2\right)^8 \times a^7$ b $r \times \left(r^3\right)^{10}$ c $\dfrac{\left(m^5\right)^3}{m^{11}}$ d $\dfrac{\left(u^2\right)^4}{u^7}$

e $\dfrac{\left(y^3\right)^3}{y}$ f $\left(x^3\right)^2 \times \left(x^3\right)^2$ g $\dfrac{\left(p^4\right)^5 \times \left(p^2\right)^8}{\left(p^3\right)^3}$ h $\dfrac{\left(d^5\right)^8}{\left(d^2\right)^2 \times \left(d^3\right)^6}$

N5 Rules 4 and 5: When the power is 0 and negative powers

Consider the first four positive powers of 10:

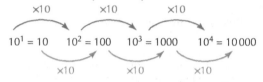

$10^1 = 10 \quad 10^2 = 100 \quad 10^3 = 1000 \quad 10^4 = 10\,000$

> **Hint** The power increases by 1 every time you multiply by 10.

The list can be extended by dividing by 10:

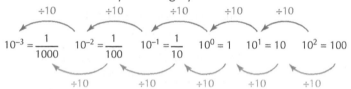

$10^{-3} = \dfrac{1}{1000} \quad 10^{-2} = \dfrac{1}{100} \quad 10^{-1} = \dfrac{1}{10} \quad 10^0 = 1 \quad 10^1 = 10 \quad 10^2 = 100$

> **Hint** The power decreases by 1 every time you divide by 10.

From this you can see that:

$$10^0 = 1 \qquad 10^{-1} = \frac{1}{10} = \frac{1}{10^1} \qquad 10^{-2} = \frac{1}{100} = \frac{1}{10^2} \qquad 10^{-3} = \frac{1}{1000} = \frac{1}{10^3}$$

So, for example, $10^{-6} = \dfrac{1}{10^6}$

A similar pattern exists with bases other than 10. For example, for base 2:

$$2^{-4} = \frac{1}{16} \quad 2^{-3} = \frac{1}{8} \quad 2^{-2} = \frac{1}{4} \quad 2^{-1} = \frac{1}{2} \quad 2^0 = 1 \quad 2^1 = 2 \quad 2^2 = 4 \quad 2^3 = 8 \quad 2^4 = 16$$

Notice:

$$2^0 = 1 \qquad 2^{-1} = \frac{1}{2^1} = \frac{1}{2} \qquad 2^{-2} = \frac{1}{2^2} = \frac{1}{4} \qquad 2^{-3} = \frac{1}{2^3} = \frac{1}{8}$$

In a similar way, $5^{-3} = \dfrac{1}{5^3}$, $6^{-5} = \dfrac{1}{6^5}$ and so on.

You can also see from these examples that $10^0 = 1$ and $2^0 = 1$.

In a similar way, $7^0 = 1$ and $9^0 = 1$ and so on.

These examples illustrate two more rules for indices:

> **Important**
>
> **Rule 4:** $a^0 = 1$ **Rule 5:** $a^{-x} = \dfrac{1}{a^x}$

N5 **Example 5.8**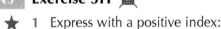

a Evaluate:

  i p^0 ii $4a^0$

b Express with a positive index:

  i a^{-7} ii $2y^{-6}$

c Express with a negative index:

  i $\dfrac{1}{b}$ ii $\dfrac{1}{5x^2}$

a i $p^0 = 1$ •——————————————————— $\boxed{\text{Rule 4 says that 'anything to the power 0' equals 1.}}$

  ii $4a^0 = 4 \times a^0$

    $= 4 \times 1 = 4$

b i $a^{-7} = \dfrac{1}{a^7}$ •——————— $\boxed{\begin{array}{l}\text{Using rule 5, to rewrite an expression with a}\\ \text{negative index as one with a positive index, write}\\ \text{the expression as the denominator of a fraction}\\ \text{which has 1 as its numerator, that is, } \dfrac{1}{\ldots}\end{array}}$

  ii $2y^{-6} = \dfrac{2}{y^6}$ •——————— $\boxed{\begin{array}{l}\text{Only the variable } y \text{ is raised to the negative index,}\\ \text{so write the 2 as the numerator of the fraction.}\end{array}}$

c i $\dfrac{1}{b} = \dfrac{1}{b^1} = b^{-1}$ •———————————————— $\boxed{\text{Using } b = b^1}$

  ii $\dfrac{1}{5x^2} = \dfrac{x^{-2}}{5} = \dfrac{1}{5}x^{-2}$ •——— $\boxed{\begin{array}{l}\text{Only the variable } x \text{ is raised to the index, so the}\\ \text{5 remains in the denominator of the fraction.}\end{array}}$

N5 **Exercise 5H**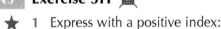

★ 1 Express with a positive index:

  a a^{-5} b x^{-9} c p^{-1} d 7^{-4}

  e 10^{-6} f $4t^{-2}$ g $8y^{-3}$ h $\dfrac{u^{-7}}{3}$

★ 2 Evaluate:

  a 8^0 b 7^0 c r^0 d y^0

  e $2b^0$ f $6m^0$ g $\dfrac{x^0}{5}$ h $\dfrac{a^0}{4}$

★ 3 Express with a negative index:

  a $\dfrac{1}{x^7}$ b $\dfrac{1}{n^8}$ c $\dfrac{1}{h}$ d $\dfrac{2}{p^3}$

  e $\dfrac{12}{e^4}$ f $\dfrac{1}{3k^2}$ g $\dfrac{1}{5v^5}$ h $\dfrac{2}{3j^{10}}$

N5 Rules 6 and 7: Fractional powers

You can use a calculator to check the following:

$$4^{\frac{1}{2}} = 2 \qquad 9^{\frac{1}{2}} = 3 \qquad 16^{\frac{1}{2}} = 4 \qquad 25^{\frac{1}{2}} = 5$$

On page 62 it was stated that:

$$\sqrt{4} = 2 \qquad \sqrt{9} = 3 \qquad \sqrt{16} = 4 \qquad \sqrt{25} = 5$$

Therefore:

> **Hint** Remember that only the positive roots are considered in this section.

$$\sqrt{4} = 4^{\frac{1}{2}} \qquad \sqrt{9} = 9^{\frac{1}{2}} \qquad \sqrt{16} = 16^{\frac{1}{2}} \qquad \sqrt{25} = 25^{\frac{1}{2}}$$

In a similar way, you can check with a calculator that:

$$\sqrt[3]{27} = 27^{\frac{1}{3}} \ (= 3) \qquad \sqrt[4]{16} = 16^{\frac{1}{4}} \ (= 2) \qquad \sqrt[6]{1\,000\,000} = 1\,000\,000^{\frac{1}{6}} \ (= 10)$$

and so on.

These examples illustrate two more rules for indices:

> **Important**
>
> **Rule 6:** $\sqrt{a} = a^{\frac{1}{2}}$ **Rule 7:** $\sqrt[x]{a} = a^{\frac{1}{x}}$

N5 Example 5.9

a **Without** the aid of a calculator, evaluate:

  i $36^{\frac{1}{2}}$ ii $8^{\frac{1}{3}}$

b Express with a fractional index:

  i $\sqrt[5]{y}$ ii $\sqrt[9]{2}$

c Express in 'root form':

  i $p^{\frac{1}{4}}$ ii $x^{\frac{1}{k}}$

a i $36^{\frac{1}{2}} = \sqrt{36}$

        $= 6$

  ii $8^{\frac{1}{3}} = \sqrt[3]{8}$

      $= 2$

b i $\sqrt[5]{y} = y^{\frac{1}{5}}$

  ii $\sqrt[9]{2} = 2^{\frac{1}{9}}$

c i $p^{\frac{1}{4}} = \sqrt[4]{p}$

  ii $x^{\frac{1}{k}} = \sqrt[k]{x}$

N5 **Exercise 5I** 🖩

★ 1 Without the aid of a calculator, evaluate:

     a $100^{\frac{1}{2}}$ **b** $49^{\frac{1}{2}}$ **c** $27^{\frac{1}{3}}$ **d** $32^{\frac{1}{5}}$

     e $16^{\frac{1}{4}}$ **f** $64^{\frac{1}{3}}$ **g** $1^{\frac{1}{5}}$ **h** $1\,000\,000^{\frac{1}{6}}$

★ 2 Express with a fractional index:

     a $\sqrt{h}$ **b** $\sqrt{b}$ **c** $\sqrt[3]{a}$ **d** $\sqrt[4]{g}$

     e $\sqrt[6]{t}$ **f** $\sqrt[k]{y}$ **g** $\sqrt{5}$ **h** $\sqrt[3]{12}$

★ 3 Express in root form:

     a $m^{\frac{1}{2}}$ **b** $t^{\frac{1}{3}}$ **c** $w^{\frac{1}{8}}$ **d** $a^{\frac{1}{y}}$

     e $p^{\frac{1}{t}}$ **f** $k^{\frac{1}{k}}$ **g** $2^{\frac{1}{3}}$ **h** $6^{\frac{1}{4}}$

N5 ## Rule 8: Non-unit fractional powers

The examples you have met so far have involved powers where the fraction is a **unit fraction.**
That is, where the numerator was 1, for example, $\frac{1}{2}$, $\frac{1}{3}$, $\frac{1}{x}$ and so on.

However, fractional powers can also involve non-unit fractions such as $\frac{2}{3}$ or $\frac{7}{4}$

Numbers raised to non-unit fractional powers, for example, $4^{\frac{3}{2}}$, $8^{\frac{4}{3}}$ and $1000^{\frac{2}{3}}$, can often be evaluated without using a calculator, using combinations of the rules of indices.

Consider $4^{\frac{3}{2}}$:

$$4^{\frac{3}{2}} = 4^{3 \times \frac{1}{2}}$$

$$= \left(4^3\right)^{\frac{1}{2}} \qquad \bullet\!\!-\!\!\!-\!\!\!-\!\!\!- \boxed{\text{Using rule 3: } (a^x)^y = a^{xy}}$$

$$= \sqrt{4^3} \qquad \bullet\!\!-\!\!\!-\!\!\!-\!\!\!- \boxed{\text{Using rule 6: } \sqrt{a} = a^{\frac{1}{2}}}$$

Similarly:

$$8^{\frac{4}{3}} = 8^{4 \times \frac{1}{3}} \qquad\qquad\qquad 1000^{\frac{2}{3}} = 1000^{2 \times \frac{1}{3}}$$

$$= \left(8^4\right)^{\frac{1}{3}} \qquad \text{and} \qquad\qquad = \left(1000^2\right)^{\frac{1}{3}}$$

$$= \sqrt[3]{8^4} \qquad\qquad\qquad\qquad = \sqrt[3]{1000^2}$$

So:

$$4^{\frac{3}{2}} = \sqrt{4^3} \qquad 8^{\frac{4}{3}} = \sqrt[3]{8^4} \qquad 1000^{\frac{2}{3}} = \sqrt[3]{1000^2}$$

In a similar way, $9^{\frac{7}{2}} = \sqrt{9^7}$ and $16^{\frac{3}{4}} = \sqrt[4]{16^3}$ and so on.

These examples illustrate the eighth rule of indices:

It is usually simpler to take the root first before raising to the power, because this makes the numbers easier to calculate, as Example 5.10 shows.

N5

Example 5.10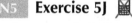

a Without the aid of a calculator, evaluate:

   i $\ 4^{\frac{3}{2}}$ ii $\ 1000^{\frac{2}{3}}$

b Express in 'root and power' form:

   i $\ t^{\frac{7}{4}}$ ii $\ x^{\frac{a}{b}}$

c Express with a fractional index:

   i $\ \sqrt[4]{a^9}$ ii $\ \sqrt[3]{r^2}$

a i $\ 4^{\frac{3}{2}} = \sqrt{4^3} = \left(\sqrt{4}\right)^3 = 2^3 = 8$

   ii $\ 1000^{\frac{2}{3}} = \sqrt[3]{1000^2} = \left(\sqrt[3]{1000}\right)^2 = 10^2 = 100$

b i $\ t^{\frac{7}{4}} = \sqrt[4]{t^7} = \left(\sqrt[4]{t}\right)^7$ ii $\ x^{\frac{a}{b}} = \sqrt[b]{x^a} = \left(\sqrt[b]{x}\right)^a$

c i $\ \sqrt[4]{a^9} = a^{\frac{9}{4}}$ ii $\ \sqrt[3]{r^2} = r^{\frac{2}{3}}$

N5

Exercise 5J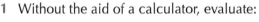

★ **1** Without the aid of a calculator, evaluate:

  a $\ 100^{\frac{3}{2}}$ b $\ 16^{\frac{3}{4}}$ c $\ 4^{\frac{5}{2}}$ d $\ 81^{\frac{3}{4}}$

  e $\ 25^{\frac{3}{2}}$ f $\ 1^{\frac{7}{2}}$ g $\ 36^{\frac{3}{2}}$ h $\ 8^{\frac{7}{3}}$

2 Express in 'root and power' form:

  a $\ a^{\frac{3}{4}}$ b $\ p^{\frac{2}{5}}$ c $\ r^{\frac{5}{4}}$ d $\ x^{\frac{11}{2}}$

  e $\ 49^{\frac{u}{v}}$ f $\ 2^{\frac{a}{b}}$ g $\ p^{\frac{a}{b}}$ h $\ x^{\frac{c}{d}}$

3 Express with a fractional index:

  a $\ \sqrt[3]{a^7}$ b $\ \sqrt[5]{x^2}$ c $\ \left(\sqrt[5]{r}\right)^6$ d $\ \left(\sqrt[6]{r}\right)^5$

  e $\ \sqrt[5]{9^3}$ f $\ \sqrt[4]{3^7}$ g $\ \left(\sqrt{5}\right)^5$ h $\ \left(\sqrt{11}\right)^9$

N5 **Combining rules**

Example 5.11 illustrates how to apply rules 1 and 2 to negative and fractional powers.

N5 **Example 5.11** ▨

a Express each of the following with a single index.

  i $p^{-3} \times p^{10}$ **ii** $a^{\frac{3}{5}} \times a^{-\frac{2}{3}}$ **iii** $\dfrac{y^{-5}}{y}$ **iv** $r^{\frac{7}{2}} \div r^{-\frac{1}{2}}$

b Simplify fully:

  i $\dfrac{6x \times 2x^{-6}}{3x^{-2}}$ **ii** $\dfrac{20a^{\frac{7}{4}}}{5a \times 2a^{\frac{1}{3}}}$

a **i** $p^{-3} \times p^{10} = p^{-3+10} = p^7$ ●————————(Using rule 1: $a^x \times a^y = a^{x+y}$)

  ii $a^{\frac{3}{5}} \times a^{-\frac{2}{3}} = a^{\frac{3}{5}+\left(-\frac{2}{3}\right)} = a^{\frac{3}{5}-\frac{2}{3}}$ ●———(When adding a negative you subtract.)

  (In order to add or subtract fractions they must have the same denominator, so find the lowest common denominator, that is, 15.)

  $= a^{\frac{9}{15}-\frac{10}{15}}$ ●

  $= a^{-\frac{1}{15}}$

  iii $\dfrac{y^{-5}}{y} = y^{-5-1} = y^{-6}$ ●

  (First, use $y = y^1$

  Then use rule 2: $\dfrac{a^x}{a^y} = a^{x-y}$)

  iv $r^{\frac{7}{2}} \div r^{-\frac{1}{2}} = r^{\frac{7}{2}-\left(-\frac{1}{2}\right)}$ ●————————————(Use rule 2.)

  $= r^{\frac{7}{2}+\frac{1}{2}}$ ●————————(When subtracting a negative you add.)

  $= r^{\frac{8}{2}} = r^4$

b **i** $\dfrac{6x \times 2x^{-6}}{3x^{-2}} = \dfrac{12x^1 \times x^{-6}}{3x^{-2}}$ ●————(First, begin to simplify the numerator.)

  $= \dfrac{12x^{1+(-6)}}{3x^{-2}} = \dfrac{12x^{1-6}}{3x^{-2}}$

  $= \dfrac{^4\cancel{12}x^{-5}}{\cancel{3}x^{-2}}$ ●————(Divide the whole numbers and the variables separately.)

  $= 4x^{-5-(-2)} = 4x^{-5+2}$

  $= 4x^{-3}$

  ii $\dfrac{20a^{\frac{7}{4}}}{5a \times 2a^{\frac{1}{3}}} = \dfrac{20a^{\frac{7}{4}}}{10a^1 \times a^{\frac{1}{3}}} = \dfrac{20a^{\frac{7}{4}}}{10a^{1+\frac{1}{3}}}$ ●———(First, begin to simplify the denominator.)

  $= \dfrac{^2\cancel{20}a^{\frac{7}{4}}}{\cancel{10}a^{\frac{4}{3}}} = 2a^{\frac{7}{4}-\frac{4}{3}}$

  $= 2a^{\frac{21}{12}-\frac{16}{12}} = 2a^{\frac{5}{12}}$

N5 **Exercise 5K**

1 Express each of the following with a single index.

a $p^8 \times p^{-1}$ b $t^5 \times t^{-12}$ c $x^{-6} \times x^{10}$ d $y^{-20} \times y$

e $a^{-3} \times a^{-4}$ f $u^{-1} \times u^{-1}$ g $w^{-5} \times w^{10} \times w^{-3}$ h $j^2 \times j \times j^{-5}$

i $4x^{-4} \times 6x^3$ j $5y^7 \times 3y^{-6}$

2 Express each of the following with a single index.

a $\dfrac{p^6}{p^{-2}}$ b $a^5 \div a^{-5}$ c $\dfrac{h^2}{h^{-4}}$ d $\dfrac{a^{-3}}{a}$

e $\dfrac{t^{-1}}{t^3}$ f $\dfrac{h^{-8}}{h^8}$ g $g^{-4} \div g^{-1}$ h $\dfrac{k^{-10}}{k^{-20}}$

i $50x^7 \div 10x^{-2}$ j $\dfrac{24m^{-4}}{8m^2}$

3 Express each of the following with a single index.

a $p^{\frac{1}{5}} \times p^{\frac{3}{5}}$ b $m^{\frac{1}{2}} \times m^{\frac{1}{2}}$ c $a^{\frac{3}{4}} \times a^{-\frac{1}{4}}$ d $w^{-\frac{2}{3}} \times w^{\frac{1}{3}}$

e $y^{-\frac{1}{3}} \times y^{-\frac{2}{3}}$ f $g^{-\frac{1}{4}} \times g^{-\frac{1}{4}}$ g $5r^{\frac{3}{4}} \times 2r^{\frac{1}{4}}$ h $9n^{\frac{5}{7}} \times 9n^{-\frac{2}{7}}$

4 Express each of the following with a single index.

a $\dfrac{a^{\frac{5}{6}}}{a^{\frac{1}{6}}}$ b $n^{\frac{3}{4}} \div n^{\frac{1}{4}}$ c $\dfrac{t^{\frac{2}{3}}}{t^{-\frac{1}{3}}}$ d $\dfrac{u^{\frac{7}{2}}}{u^{-\frac{1}{2}}}$

e $y^{-\frac{4}{5}} \div y^{\frac{1}{5}}$ f $\dfrac{v^{-\frac{1}{5}}}{v^{\frac{1}{5}}}$ g $a^{-\frac{1}{7}} \div a^{-\frac{2}{7}}$ h $\dfrac{x^{-\frac{2}{3}}}{x^{-\frac{2}{3}}}$

i $\dfrac{16a^{\frac{7}{9}}}{4a^{\frac{2}{9}}}$ j $6h^{-\frac{1}{4}} \div 6h^{\frac{3}{4}}$

★ 5 Simplify:

a $\dfrac{m^3}{m^{\frac{1}{2}}}$ b $a \times a^{\frac{1}{2}}$ c $p^{-\frac{1}{2}} \div p$ d $p^2 \times p^{-\frac{1}{2}}$

e $3x^{-\frac{3}{4}} \times 4x^2$ f $30d^{\frac{1}{3}} \div 6d$ g $\dfrac{12y^2}{4y^{\frac{1}{2}}}$ h $10y^3 \times 2y^{\frac{1}{3}}$

6 Simplify fully:

a $\dfrac{4x^{-3} \times 5x^2}{10x^3}$ b $\dfrac{16p^{-1}}{4p \times 2p^{-4}}$ c $\dfrac{20a^{\frac{1}{3}} \times 3a^{\frac{2}{3}}}{5a}$ d $\dfrac{40r^{\frac{5}{3}}}{2r \times 2r^{\frac{1}{2}}}$

Important

Remember these rules for working with surds:

Rule 1: If x and y are whole numbers then: $\sqrt{x}\sqrt{y} = \sqrt{xy}$
Rule 2: If x and y are whole numbers then: $\dfrac{\sqrt{x}}{\sqrt{y}} = \sqrt{\dfrac{x}{y}}$

Remember these rules for working with indices:

Rule 1: $a^x \times a^y = a^{x+y}$	**Rule 5:** $a^{-x} = \dfrac{1}{a^x}$
Rule 2: $\dfrac{a^x}{a^y} = a^{x-y}$	**Rule 6:** $\sqrt{a} = a^{\frac{1}{2}}$
Rule 3: $\left(a^x\right)^y = a^{xy}$	**Rule 7:** $\sqrt[x]{a} = a^{\frac{1}{x}}$
Rule 4: $a^0 = 1$	**Rule 8:** $a^{\frac{x}{y}} = \sqrt[y]{a^x}$

N5 **Chapter 5 review** ⧗

1 Express as a single surd:

$\sqrt{3} \times \sqrt{5}$

2 Express as a product of two surds:

$\sqrt{2 \times 13}$

3 Simplify:

a $\sqrt{75}$ b $\sqrt{28} + 5\sqrt{7} - \sqrt{63}$

4 Evaluate:

a $\sqrt{\dfrac{49}{81}}$ b $\dfrac{\sqrt{1500}}{\sqrt{15}}$

5 Evaluate:

a 5^0 b 10^{-2} c $16^{\frac{1}{2}}$ d $81^{\frac{1}{4}}$ e $1000^{\frac{2}{3}}$

6 Simplify:

a $p^6 \times p^4$ b $\dfrac{t^4}{t^{-1}}$ c $\left(u^{10}\right)^3$ d $2t^{\frac{3}{4}} \times 6t^{-\frac{1}{4}}$

e $\dfrac{18a^{-\frac{2}{3}}}{6a^{\frac{1}{3}}}$ f $\dfrac{u^5 \times u^{-3}}{u^2}$ g $\dfrac{24y^{\frac{5}{2}}}{3y^{\frac{3}{2}} \times 2y^{-\frac{1}{2}}}$

- I can recall and apply the rule $\sqrt{x}\sqrt{y} = \sqrt{xy}$ ★ Exercise 5A Q1
  ★ Exercise 5B Q1, Q3
- I can recall and apply the rule $\dfrac{\sqrt{x}}{\sqrt{y}} = \sqrt{\dfrac{x}{y}}$ ★ Exercise 5D Q2
- I can recall and apply the rule $a^x \times a^y = a^{x+y}$ ★ Exercise 5E Q1
- I can recall and apply the rule $\dfrac{a^x}{a^y} = a^{x-y}$ ★ Exercise 5F Q2
- I can recall and apply the rule $\left(a^x\right)^y = a^{xy}$ ★ Exercise 5G Q1
- I can recall and apply the rule $a^0 = 1$ ★ Exercise 5H Q2
- I can recall and apply the rule $a^{-x} = \dfrac{1}{a^x}$ ★ Exercise 5H Q1, Q3
- I can recall and apply the rule $\sqrt{a} = a^{\frac{1}{2}}$ ★ Exercise 5I Q1
- I can recall and apply the rule $\sqrt[x]{a} = a^{\frac{1}{x}}$ ★ Exercise 5I Q2, Q3
- I can recall and apply the rule $a^{\frac{x}{y}} = \sqrt[y]{a^x}$ ★ Exercise 5J Q1
- I can apply combinations of rules to simplify expressions.
  ★ Exercise 5K Q5

6 Algebra 2

This chapter will show you how to:

- solve linear equations
- change the subject of a formula
- solve a quadratic equation from its factorised form
- solve a quadratic equation by factorising
- solve a quadratic equation using the quadratic formula.

You should already know how to:

- expand a pair of brackets with a number at the front
- factorise an expression by taking out a numerical common factor
- simplify expressions with one or more variables (or letters)
- work with algebraic expressions involving expansion of brackets (see pages 14–16)
- factorise an algebraic expression (involving a common factor, difference of two squares or trinomial) (see pages 16–21)
- use the distributive law in an expression with a numerical common factor to produce a sum of terms.

N4 Solving simple linear equations

A **simple linear equation** is one which contains numbers and a variable (letter) – such as x or t – but which doesn't contain anything more complicated such as x^2 or t^3, etc. To **solve** such an equation means to find the number which the variable represents. To solve for x, you need to work through the solution in order to obtain an expression '$x = ...$', that is, get x on its own, usually to the left of the equals sign as illustrated by Examples 6.1 and 6.2.

N4 Example 6.1

Solve the following equations.

a $x + 5 = 7$ b $p - 1 = 12$ c $4a = 22$ d $\frac{t}{3} = 4$

a $x + 5 = 7$

$x + 5 - 5 = 7 - 5$ To get x on its own and 'undo' $+ 5$ on the LHS, you need to subtract 5 from both sides.

$x = 2$

Check:

LHS $= x + 5 = 2 + 5 = 7 =$ RHS ✓   Check that your solution is correct by substituting it back into the original equation. Your solution is correct if LHS = RHS.

b $p - 1 = 12$

$p - 1 + 1 = 12 + 1$

$p = 13$ To undo -1 on the LHS, add 1 to both sides.

c $\quad 4a = 22$

$$\frac{4a}{4} = \frac{22}{4}$$ •————— To undo × 4 on the LHS, divide both sides by 4.

$$a = \frac{^{11}\cancel{22}}{\cancel{4}_{\,2}}$$

$$= \frac{11}{2}$$

$$a = 5\tfrac{1}{2} \text{ or } 5{\cdot}5$$ •————— Express the improper fraction as a mixed number or as a decimal fraction.

d $\quad \dfrac{t}{3} = 4$

$$\frac{t}{3} \times 3 = 4 \times 3$$ •————— To undo ÷3 on the LHS, multiply both sides by 3.

$$t = 12$$

N4 ## Example 6.2

Solve the following equations.

a $\quad 2a + 9 = 11$ $\qquad$ b $\quad 17 = 3x - 4$ $\qquad$ c $\quad \dfrac{1}{2}t - 1 = 6$

a $\quad 2a + 9 = 11$

$$2a + 9 - 9 = 11 - 9$$ •————— When solving an equation aim to get the variable alone. Start by subtracting 9 from both sides.

$$2a = 2$$

$$\frac{2a}{2} = \frac{2}{2}$$ •————— Then divide both sides by 2.

$$a = 1$$

b $\quad 17 = 3x - 4$

$$3x - 4 = 17$$ •————— Swap the LHS and RHS so that the variable is in the more familiar position, on the LHS.

$$3x - 4 + 4 = 17 + 4$$

$$3x = 21$$

$$\frac{3x}{3} = \frac{21}{3}$$

$$x = 7$$

c $\quad \dfrac{1}{2}t - 1 = 6$

$$\frac{1}{2}t - 1 + 1 = 6 + 1$$

$$\frac{1}{2}t = 7$$

$$\frac{1}{2}t \times 2 = 7 \times 2$$ •————— To undo $\times \frac{1}{2}$ multiply both sides of the equation by 2.

$$t = 14$$

N4 **Exercise 6A**

1 Solve the following equations.

   a $x + 2 = 9$ **b** $t + 5 = 6$ **c** $y + 11 = 23$ **d** $m + 150 = 230$

   e $a + 8 = 8$ **f** $a - 4 = 6$ **g** $m - 3 = 10$ **h** $y - 9 = 0$

   i $w - 5 = 2$ **j** $x - 200 = 50$

2 Solve the following equations.

   a $5a = 45$ **b** $10p = 60$ **c** $3w = 12$ **d** $8b = 40$

   e $2x = 2$ **f** $5y = 55$ **g** $\frac{x}{2} = 9$ **h** $\frac{t}{5} = 3$

   i $\frac{a}{7} = 2$ **j** $\frac{y}{4} = 1$

3 Solve the following equations.

   a $x + 2 \cdot 5 = 5 \cdot 5$ **b** $m + 2 \cdot 4 = 6$ **c** $r + 6 = 7 \cdot 2$

   d $b - 4 \cdot 5 = 20$ **e** $y - 2 = 6 \cdot 8$ **f** $p - 10 \cdot 5 = 1 \cdot 5$

4 Solve the following equations.

   a $4t = 10$ **b** $3x = 7 \cdot 8$ **c** $10p = 24$

   d $\frac{t}{2} = 3 \cdot 5$ **e** $\frac{m}{3} = 2 \cdot 2$ **f** $\frac{y}{10} = 5 \cdot 4$

5 Solve the following equations.

   a $2x + 5 = 17$ **b** $3t + 4 = 16$ **c** $10p + 9 = 59$ **d** $9 = 4t + 5$

   e $3m - 2 = 4$ **f** $5w - 1 = 14$ **g** $6t - 6 = 0$ **h** $1 = 3x - 2$

> **Hint** If the variable is on the RHS of the equation, you could find it easier to swap the expressions on the LHS and RHS so that the variable is on the LHS. But be careful to make sure you don't accidentally change anything in either expression.

6 Solve the following equations.

   a $\frac{1}{3}t + 4 = 5$ **b** $\frac{1}{2}x + 3 = 5$ **c** $\frac{1}{4}p - 1 = 5$

   d $9 = \frac{1}{2}a - 1$ **e** $\frac{t}{2} + 7 = 9$ **f** $5 = \frac{w}{5} + 4$

7 Solve the following equations.

   a $7t + 2 = 16$ **b** $11 = 11 + 5y$ **c** $\frac{r}{2} - 3 = 5$

   d $3x - 3 = 6$ **e** $10 = \frac{u}{3} + 3$ **f** $7 = 2a - 3$

N4 ▶N5 Solving equations with an unknown on both sides

In many equations, there is an unknown variable (or letter) on both sides of the equation. You need to work through a solution so that you get an equation with the variable on only one side. Example 6.3 shows how to solve such equations.

N4 ▶N5 Example 6.3

Solve the following equations algebraically.

a $6x + 5 = x + 60$ b $7a - 6 = 8 + 5a$ c $21 - p = 6p - 7$

a $6x + 5 = x + 60$

$6x + 5 - 5 = x + 60 - 5$ ●————————————(Undo + 5.)

$6x = x + 55$

$6x - x = x + 55 - x$ ●————(To get an equation with the variable on only one side, subtract x from both sides. Remember that $x = 1x$.)

$5x = 55$

$\dfrac{5x}{5} = \dfrac{55}{5}$

$x = 11$

b $7a - 6 = 8 + 5a$

$7a - 6 + 6 = 8 + 5a + 6$

$7a = 14 + 5a$

$7a - 5a = 14 + 5a - 5a$ ●————————————(Subtract $5a$ from both sides.)

$2a = 14$

$\dfrac{2a}{2} = \dfrac{14}{2}$

$a = 7$

c $21 - p = 6p - 7$

$6p - 7 = 21 - p$ ●————————————(Swap LHS and RHS.)

$6p - 7 + 7 = 21 - p + 7$

$6p = 28 - p$

$6p + p = 28 - p + p$

$7p = 28$

$\dfrac{7p}{7} = \dfrac{28}{7}$

$p = 4$

N4 Exercise 6B

1 Solve the following equations algebraically.

a $6x + 1 = 2x + 21$ b $7t + 4 = 4t + 34$ c $10p + 1 = p + 28$

d $5m + 5 = 4m + 12$ e $2y + 18 = 9y + 11$ f $14 + w = 4w + 8$

(continued)

g $9p - 2 = 6p + 4$ h $7t - 1 = 2t + 14$ i $6t - 1 = 2t + 3$

j $8a - 2 = a + 19$ k $8 + f = 4f - 1$ l $57 + 2q = 10q - 7$

2 Solve the following equations.

a $4x + 10 = 20 - x$ b $8t + 5 = 35 - 2t$ c $4p + 3 = 21 - 5p$

d $10 - 4y = 5y + 1$ e $8 - y = 4y + 8$ f $14 - 7y = 5y + 2$

N4 Solving linear equations with brackets

▶N5 Chapter 2 showed how to expand pairs of brackets by multiplying. You can also solve linear equations which contain one or more pairs of brackets. The first thing to do is expand any brackets, as shown in Example 6.4.

N4
▶N5

Example 6.4

Solve the following equations.

a $6(2x + 1) = 30$ b $3(3t - 2) + 5 = 8$ c $4(20 - a) = 25 + 3(2a + 5)$

a $6(2x + 1) = 30$

$12x + 6 = 30$ ● ————— First, expand the brackets by multiplying both $2x$ **and** 1 by 6.

$12x + 6 - 6 = 30 - 6$ ● ————— Now solve as before, by undoing any additions or subtractions, and then any multiplications or divisions.

$12x = 24$

$\dfrac{12x}{12} = \dfrac{24}{12}$

$x = 2$

b $3(3t - 2) + 5 = 8$

$9t - 6 + 5 = 8$ ● ————— Expand the brackets.

$9t - 1 = 8$ ● ————— Collect like terms.

$9t - 1 + 1 = 8 + 1$

$9t = 9$

$\dfrac{9t}{9} = \dfrac{9}{9}$

$t = 1$

c $4(20 - a) = 25 + 3(2a + 5)$

$80 - 4a = 25 + 6a + 15$ ● ————— Expand the brackets on **both** sides of the equation.

$80 - 4a = 40 + 6a$ ● ————— Collect like terms.

$80 - 4a + 4a = 40 + 6a + 4a$

$80 = 40 + 10a$

$80 - 40 = 40 - 40 + 10a$

$40 = 10a$

$\dfrac{40}{10} = \dfrac{10a}{10}$

$a = 4$

Exercise 6C

1 Solve the following equations.

a $5(x + 2) = 20$ b $8(t + 1) = 32$ c $12 = 6(m - 3)$

d $10 = 10(p - 5)$ e $2(t + 3) + 5 = 13$ f $8(k - 3) + 6 = 22$

g $29 = 8(4 + a) - 3$ h $3(1 + m) - 2 = 13$ i $22 = 7(t + 2) - 6$

2 Solve the following equations.

a $4(2h + 5) + 1 = 29$ b $3(3x + 1) + 1 = 31$ c $4(4m + 3) - 5 = 39$

d $53 = 5(3h + 2) - 2$ e $2(5g - 2) + 5 = 21$ f $35 = 4(3x - 1) + 3$

g $1 = 2(4j - 3) - 1$ h $10(2a - 2) - 15 = 5$ i $22 = 5(1 + t) - 18$

3 Solve the following equations algebraically.

a $6(2x + 1) = 5(x + 4)$ b $4(3t + 3) = 3(2t + 12)$

c $5(3m - 2) = 4(2m + 1)$ d $3(2m + 17) = 5(4m - 1)$

e $2(5h + 2) + 4 = 3(2h + 4)$ f $4(3x + 1) + 6 = 5(3x - 4)$

g $5(2p + 2) - 30 = 2(11 - 2p)$ h $2(20 - y) = 1 + 3(3y + 2)$

Solving equations with negative or fractional solutions

Sometimes equations have solutions which are negative and/or fractions, as shown in Example 6.5.

Example 6.5

Solve the following equations.

a $4(2t + 3) = 8$ b $10 - x = 4x + 25$

a $4(2t + 3) = 8$

$8t + 12 = 8$ Expand the brackets.

$8t + 12 - 12 = 8 - 12$

$8t = -4$

$\dfrac{8t}{8} = \dfrac{-4}{8}$

$t = -\dfrac{4}{8} = -\dfrac{\overset{1}{\cancel{4}}}{\underset{2}{\cancel{8}}}$ Simplify the fraction.

$= -\dfrac{1}{2}$

b $10 - x = 4x + 25$

   $10 - x + x = 4x + 25 + x$

   $10 = 5x + 25$

   $10 - 25 = 5x + 25 - 25$

   $-15 = 5x$

   $5x = -15$

   $\dfrac{5x}{5} = \dfrac{-15}{5}$

   $x = -3$

N4

N5 **Exercise 6D**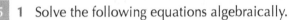

1 Solve the following equations algebraically.

   a $2x + 3 = 4$ **b** $3m - 1 = 1$ **c** $5t + 13 = 3$

   d $1 = 4t + 21$ **e** $5a + 4 = 2$ **f** $3 = 14t + 10$

★ **2** Solve the following equations.

   a $5x + 9 = x + 6$ **b** $20 + 2m = 5m + 23$ **c** $4y - 7 = 5 - y$

   d $6(2h + 1) + 2 = 2$ **e** $1 = 2(4g - 3) + 15$ **f** $4(3x + 1) - 3 = 5$

3 Solve the following equations algebraically.

   a $5(2x + 8) = 2(x + 4)$ **b** $2(3t + 1) = 3(6t + 2)$

   c $5(3m - 3) = 10(2m - 1)$ **d** $3(2m + 1) - 3 = 5(4m - 1) - 2$

N4 ## Changing the subject of a formula

N5 A **formula** states a relationship or rule between a set of variables. For example, the formula for the area of a circle is $A = \pi r^2$, where A is the area and r is the radius. So if you know the radius of a circle, you can calculate the area. If you know the area but want to find the radius, you need to change the subject of the formula (the variable on the LHS of the equals sign) to $r = \ldots$

N4

N5 **Example 6.6**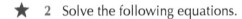

Change the subject of each formula to the variable in square brackets.

a $y = 4x - 3$ $[x]$ **b** $L = c - pt$ $[t]$ **c** $k = \dfrac{1}{2}m + 2t$ $[m]$

a $y = 4x - 3$ $[x]$ ●———————— To change the subject of this formula to x means you need to obtain a final equation beginning $x = \ldots$

   $4x - 3 = y$ ●———————— Swap LHS and RHS to get the variable x on the LHS.

   $4x - 3 + 3 = y + 3$

   $4x = y + 3$

   $\dfrac{4x}{4} = \dfrac{y + 3}{4}$

   $x = \dfrac{y + 3}{4}$ ●———— x is now the subject and the equation is of the form needed, that is, $x = \ldots$

b $L = c - pt$ $[t]$ — To make t the subject, you need to get a final equation beginning $t = ...$

$c - pt = L$ — Swap LHS and RHS.

$c - pt - c = L - c$ — Subtract c from both sides, since c means $+c$.

$-pt = L - c$

$\dfrac{-pt}{-p} = \dfrac{L - c}{-p}$ — Divide both sides of the equation by $-p$.

$t = \dfrac{L - c}{-p}$

$= \dfrac{c - L}{p}$ — Be careful when dividing by a negative.

c $k = \dfrac{1}{2}m + 2t$ $[m]$

$\dfrac{1}{2}m + 2t = k$

$\dfrac{1}{2}m + 2t - 2t = k - 2t$

$\dfrac{1}{2}m = k - 2t$

$\dfrac{1}{2}m \times 2 = (k - 2t) \times 2$ — Undo $\times \dfrac{1}{2}$. Make sure you multiply all of the RHS by 2.

$m = 2k - 4t$ — This can also be expressed as $m = 2(k - 2t)$.

N4 Exercise 6E

N5 **1** Change the subject of each formula to the variable in square brackets.

a $y = 2x + 1$ $[x]$
b $y = 5x - 2$ $[x]$
c $u = 4 + 3v$ $[v]$

d $a = 6b - 5$ $[b]$
e $y = 7 - 3x$ $[x]$
f $T = -4S + 9$ $[S]$

★ **2** Change the subject of each formula to the variable in square brackets.

a $P = 4Q + R$ $[R]$
b $P = 4Q + R$ $[Q]$
c $A = 3B - 4C$ $[B]$

d $a = bc + d$ $[c]$
e $p = 2q + rs$ $[r]$
f $E = 5F - GH$ $[H]$

3 Change the subject of each formula to the variable in square brackets.

a $g = \dfrac{1}{2}h - 4$ $[h]$
b $W = \dfrac{1}{3}T + 2R$ $[T]$
c $P = 10Q - \dfrac{1}{4}R$ $[R]$

d $y = 8 - \dfrac{1}{5}x$ $[x]$
e $R = 6U - \dfrac{1}{10}V$ $[V]$
f $a = 6b + \dfrac{1}{6}c + 3$ $[c]$

▶N5 Solving quadratic equations

A **quadratic equation** is a trinomial expression that is set equal to zero, for example, $x^2 - x - 6 = 0$. Expressions such as $x^2 - x - 6$ are called **trinomials** because they consist of **three** parts (an x^2 term, an x term and a constant or number).

This section shows how to solve quadratic equations:

- solving equations that are already factorised
- solving equations by factorising first
- solving equations using the quadratic formula.

N5 Solving factorised quadratic equations

To **solve** an equation means to find a value or values for the variable which makes the equation correct. So, solving a factorised quadratic equation, for example, $(x + 2)(x + 3) = 0$, means finding what value or values of x makes the left-hand side of the equation equal to zero.

$(x + 2)(x + 3)$ is mathematical shorthand for $(x + 2) \times (x + 3)$. So, the two expressions, $(x + 2)$ and $(x + 3)$, multiply to give zero. And if two expressions multiply to give zero, then at least one of them must be equal to zero. That is, either $(x + 2) = 0$ or $(x + 3) = 0$, or both. Example 6.7 shows how to solve such factorised equations.

N5 Example 6.7 🖩

Solve these equations.

a $(x + 2)(x + 3) = 0$ b $(5t - 3)(2t + 1) = 0$ c $2x(x - 4) = 0$

a $(x + 2)(x + 3) = 0$

  Either:

$(x + 2) = 0$ or $(x + 3) = 0$ •──[Two expressions multiplied together give an answer of zero, so at least one of them must equal zero.]

$x + 2 - 2 = 0 - 2$ or $x + 3 - 3 = 0 - 3$

$x = -2$ or $x = -3$

b $(5t - 3)(2t + 1) = 0$

  Either:

$(5t - 3) = 0$ or $(2t + 1) = 0$ •──[Either $(5t - 3)$ or $(2t + 1)$ must equal zero.]

$5t - 3 + 3 = 0 + 3$ or $2t + 1 - 1 = 0 - 1$

$5t = 3$ or $2t = -1$

$\dfrac{5t}{5} = \dfrac{3}{5}$ or $\dfrac{2t}{2} = \dfrac{-1}{2}$

$t = \dfrac{3}{5}$ or $t = -\dfrac{1}{2}$

c $2x(x - 4) = 0$

  Either:

$2x = 0$ or $(x - 4) = 0$ •──────[Either $2x$ or $(x - 4)$ must equal zero.]

$\dfrac{2x}{2} = \dfrac{0}{2}$ or $x - 4 + 4 = 0 + 4$

$x = 0$ or $x = 4$

N5 **Exercise 6F** 🖩

1 Solve these equations.

   a $(x + 1)(x + 3) = 0$ b $(t + 2)(t + 6) = 0$ c $(a + 4)(a + 10) = 0$

   d $(x + 8)(x - 2) = 0$ e $(p + 5)(p - 5) = 0$ f $(t - 1)(t + 9) = 0$

   g $(x - 4)(x + 3) = 0$ h $(m - 3)(m - 2) = 0$ i $(q - 4)(q - 7) = 0$

★ 2 Solve these equations.

   a $(x + 3)(2x + 1) = 0$ b $(3p + 2)(p + 2) = 0$ c $(t - 2)(2t - 1) = 0$

   d $(5y + 3)(y - 2) = 0$ e $(a - 5)(3a + 1) = 0$ f $(4k - 1)(2k - 9) = 0$

   g $(5x - 4)(2x + 3) = 0$ h $(7f + 3)(f + 2) = 0$ i $(2b - 3)(2b - 7) = 0$

3 Solve these equations.

   a $x(x - 3) = 0$ b $p(p + 8) = 0$ c $3t(t + 9) = 0$

   d $2a(a - 1) = 0$ e $5p(2p - 1) = 0$ f $6y(6y - 5) = 0$

N5 Solving quadratic equations by factorising first

N5 **Example 6.8** 🖩

Solve these equations.

a $x^2 + 5x + 4 = 0$ b $a^2 - a - 12 = 0$ c $t^2 - 9t + 20 = 0$

a $x^2 + 5x + 4 = 0$

   $(x + 1)(x + 4) = 0$ ●————————————————⟨ First, factorise $x^2 + 5x + 4$. ⟩

   Either:

     $(x + 1) = 0$ or $(x + 4) = 0$

  $x + 1 - 1 = 0 - 1$ or $x + 4 - 4 = 0 - 4$

       $x = -1$ or $x = -4$

b $a^2 - a - 12 = 0$

   $(a + 3)(a - 4) = 0$ ●————————————————⟨ First, factorise $a^2 - a - 12$. ⟩

   Either:

     $(a + 3) = 0$ or $(a - 4) = 0$

  $a + 3 - 3 = 0 - 3$ or $a - 4 + 4 = 0 + 4$

       $a = -3$ or $a = 4$

c $t^2 - 9t + 20 = 0$

   $(t - 4)(t - 5) = 0$

   Either:

     $(t - 4) = 0$ or $(t - 5) = 0$

  $t - 4 + 4 = 0 + 4$ or $t - 5 + 5 = 0 + 5$

       $t = 4$ or $t = 5$

Exercise 6G

1   Solve these equations by factorising.

   a $x^2 + 6x + 5 = 0$ **b** $t^2 + 4t + 3 = 0$ **c** $a^2 + 3a + 2 = 0$

   d $m^2 + 5m + 6 = 0$ **e** $p^2 + 7p + 12 = 0$ **f** $y^2 + 10y + 21 = 0$

   g $k^2 + 22k + 21 = 0$ **h** $x^2 + 8x + 7 = 0$ **i** $h^2 + 7h + 10 = 0$

★ 2   Solve these equations.

   a $x^2 + 6x - 7 = 0$ **b** $x^2 - 6x - 7 = 0$ **c** $t^2 - 4t - 5 = 0$

   d $a^2 + 2a - 3 = 0$ **e** $k^2 + k - 12 = 0$ **f** $a^2 - 8a - 20 = 0$

   g $x^2 + 7x - 8 = 0$ **h** $p^2 - 3p - 10 = 0$ **i** $x^2 + 3x - 40 = 0$

3   Solve these equations.

   a $a^2 - 4a + 3 = 0$ **b** $t^2 - 3t + 2 = 0$ **c** $x^2 - 12x + 11 = 0$

   d $y^2 - 6y + 5 = 0$ **e** $x^2 - 6x + 8 = 0$ **f** $p^2 - 8p + 15 = 0$

   g $m^2 - 12m + 20 = 0$ **h** $r^2 - 8r + 12 = 0$ **i** $x^2 - 7x + 6 = 0$

Solving quadratic equations using the quadratic formula

All of the trinomials that were factorised in Chapter 2 and all of the quadratic equations that were solved in the previous exercise had **1** as the **leading coefficient** (the number at the front) of the squared term, for example, $1x^2 + 5x + 6$, $1t^2 - t - 12$, etc. (Remember that, by convention, you don't normally write the 1 in front of x^2, t^2, etc.)

In some quadratic equations, however, either the leading coefficient is not 1 or the expression doesn't factorise easily. In such cases, the equation is solved by using the quadratic formula.

> **Important**
>
> The **roots** of the quadratic equation $ax^2 + bx + c = 0$ are given by the **quadratic formula:**
>
> $$x = \frac{-b \pm \sqrt{b^2 - 4ac}}{2a}$$
>
> The roots of a quadratic equation are another name for the two solutions to the equation.

The ± sign in the numerator of the formula indicates there are two solutions to be found. Example 6.9 shows how to apply the quadratic formula and how to deal with the ± sign.

> **Hint**   The quadratic formula is given in the Formula list in the external National 5 exam, but through practice you can memorise it too.

N5

Example 6.9

 a Without using a calculator, use the quadratic formula to find the roots of this equation:

$$2x^2 + 11x + 5 = 0$$

 b Using a calculator, use the quadratic formula to find the roots of these equations:

    i $3x^2 + 2x - 4 = 0$ to 1 decimal place   **ii** $t^2 - 3t - 2 = 0$ to 3 significant figures.

a $2x^2 + 11x + 5 = 0$

$a = 2, b = 11, c = 5$

> Compare $2x^2 + 11x + 5 = 0$ with the general quadratic equation $ax^2 + bx + c = 0$. Write down the values of a, b and c.

$$x = \frac{-b \pm \sqrt{b^2 - 4ac}}{2a}$$

> Write down the quadratic formula.

$$= \frac{-11 \pm \sqrt{11^2 - 4 \times 2 \times 5}}{2 \times 2}$$

> Substitute the values of a, b and c into the formula. Remember that $4ac$ means $4 \times a \times c$.

$$= \frac{-11 \pm \sqrt{81}}{4}$$

> Simplify.

$$x = \frac{-11 + \sqrt{81}}{4} \quad \text{or} \quad x = \frac{-11 - \sqrt{81}}{4}$$

> Replace the ± sign with separate + and − signs in two different equations in order to obtain two solutions.

$$x = \frac{-11 + 9}{4} \quad \text{or} \quad x = \frac{-11 - 9}{4}$$

$$x = \frac{-2}{4} \quad \text{or} \quad x = \frac{-20}{4}$$

$$x = -\frac{1}{2} \quad \text{or} \quad x = -5$$

b **i** $3x^2 + 2x - 4 = 0$

$a = 3, b = 2, c = -4$

> Compare the given equation with the quadratic formula. Write down the values of a, b and c. Note that in this example, c is negative.

$$x = \frac{-b \pm \sqrt{b^2 - 4ac}}{2a}$$

> Write down the quadratic formula.

$$= \frac{-2 \pm \sqrt{2^2 - 4 \times 3 \times (-4)}}{2 \times 3}$$

> Substitute the values of a, b and c.

$$= \frac{-2 \pm \sqrt{52}}{6}$$

> Simplify.

$$x = \frac{-2 + \sqrt{52}}{6} \quad \text{or} \quad x = \frac{-2 - \sqrt{52}}{6}$$

> Replace the ± sign with separate + and − signs in two different equations.

$$x = 0.868\ldots \quad \text{or} \quad x = -1.535\ldots$$

> Be careful using your calculator and write answers to at least 3 d.p.

$$x = 0.9 \quad \text{or} \quad x = -1.5 \ (1 \text{ d.p.})$$

> Round your answers to 1 d.p. as specified in the question.

ii $t^2 - 3t - 2 = 0$

$a = 1, b = -3, c = -2$ ————————————— $\boxed{\text{Write down the values of } a, b \text{ and } c.}$

$t = \dfrac{-b \pm \sqrt{b^2 - 4ac}}{2a}$ ————————— $\boxed{\text{Write down the quadratic formula in terms of } t, \text{ not } x.}$

$= \dfrac{3 \pm \sqrt{(-3)^2 - 4 \times 1 \times (-2)}}{2 \times 1}$ ———— $\boxed{\begin{array}{l}\text{Substitute the values of } a, b \text{ and } c.\\ \text{Note that if } b = -3, \text{ then } -b = 3.\end{array}}$

$= \dfrac{3 \pm \sqrt{17}}{2}$ ————————————————— $\boxed{\text{Simplify.}}$

$t = \dfrac{3 + \sqrt{17}}{2}$ or $t = \dfrac{3 - \sqrt{17}}{2}$ $\boxed{\begin{array}{l}\text{Be careful using your calculator and write}\\ \text{answers to more significant figures than needed.}\end{array}}$

$t = 3 \cdot 5615...$ or $t = -0 \cdot 5615...$ ——— $\boxed{\begin{array}{l}\text{Note that a leading zero (in front of a decimal}\\ \text{point) is \textbf{not} a significant figure and is simply a}\\ \text{place holder.}\end{array}}$

$t = 3 \cdot 56$ or $t = -0 \cdot 562$ ——————— $\boxed{\begin{array}{l}\text{Round your answers to 3 s.f.}\\ \text{as specified in the question.}\end{array}}$

N5 ## Exercise 6H

 1 Without using a calculator, use the quadratic formula to find the roots of these equations.

a $2x^2 + 9x + 4 = 0$ b $3x^2 + 4x + 1 = 0$

c $5t^2 + 17t + 6 = 0$ d $2x^2 + 9x - 5 = 0$

e $3y^2 - 4y + 1 = 0$ f $4x^2 - 7x - 2 = 0$

 2 Using a calculator, use the quadratic formula to find the roots of these equations to 1 decimal place.

★

a $2x^2 + 7x + 1 = 0$ b $t^2 + 5t + 2 = 0$

c $2x^2 + 8x + 3 = 0$ d $3y^2 + y - 1 = 0$

e $5x^2 - 9x + 2 = 0$ f $p^2 - 2p - 1 = 0$

3 Using a calculator, use the quadratic formula to find the roots of these equations to 3 significant figures.

a $x^2 + 4x + 2 = 0$ b $3t^2 + 5t + 1 = 0$

c $2y^2 + 11y + 3 = 0$ d $4x^2 + 4x - 5 = 0$

e $2m^2 - 8m + 1 = 0$ f $p^2 - p - 1 = 0$

Chapter 6 review

1 Solve algebraically.

   a $4(3a + 2) - 7 = 2(5a + 1)$ b $20 - x = 4x + 25$

2 Change the subject of the formula to the variable in square brackets.

   a $y = 1 - 2t$ $[t]$ b $P = \frac{1}{3}Q + 4R$ $[Q]$

3 Solve the equations.

   a $(x - 5)(2x + 1) = 0$ b $4t(t - 6) = 0$

4 Solve by factorising.

   a $y^2 + 8y + 7 = 0$ b $m^2 - 5m - 6 = 0$ c $t^2 - 9t + 20 = 0$

5 Use the quadratic formula $x = \dfrac{-b \pm \sqrt{b^2 - 4ac}}{2a}$ to solve the following equations.

   a $3x^2 + 7x + 2 = 0$ without using a calculator

   b $x^2 + 5x - 1 = 0$ using a calculator, rounding the roots to **1 decimal place**

   c $5y^2 - y - 2 = 0$ using a calculator, rounding the roots to **3 significant figures**.

- • I can solve linear equations. ★ Exercise 6D Q2

- • I can change the subject of a formula. ★ Exercise 6E Q2

- • I can solve a quadratic equation which has been factorised. ★ Exercise 6F Q2

- • I can solve a quadratic equation by factorising. ★ Exercise 6G Q2

- • I can solve a quadratic equation using the quadratic formula. ★ Exercise 6H Q2

7 Geometry 2

This chapter will show you how to:

- calculate the gradient of a straight line from horizontal and vertical distances
- draw a straight-line graph given its equation
- identify the gradient and y-intercept from the equation of a straight line
- find the gradient of a straight line using the formula $m = \dfrac{y_2 - y_1}{x_2 - x_1}$
- use the formula $y - b = m(x - a)$ to find the equation of a straight line given the gradient and a point on the line or two points on the line
- use and apply function notation $f(x)$.

You should already know:

- how to plot points on a coordinate diagram
- the terms origin, x-axis, y-axis, x-coordinate and y-coordinate
- the relationship between fractions, decimal fractions and division
- the correct order of arithmetic operations
- how to evaluate a formula by substitution
- how to change the subject of a formula (see Chapter 6 pages 88–89).

N4 Calculating the gradient of a straight line using horizontal and vertical distances

The **gradient** of a straight line is a measure of its steepness – the greater the gradient, the steeper the slope.

The gradient measures how much a line goes up (or down) for every 1 unit you move along (left or right).

For example, a gradient of 3 means for every 1 unit you move to the right horizontally, you move 3 vertically up.

Movement up or right is in a positive direction, whereas movement down or left is negative.

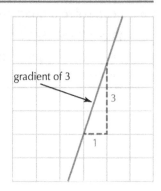

Important

The rule for calculating gradient is:

$$\text{gradient} = \frac{\text{vertical distance}}{\text{horizontal distance}}$$

N4 **Example 7.1**

Calculate the gradient of the slope in the diagram.

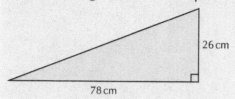

26 cm

78 cm

$\text{gradient} = \dfrac{\text{vertical distance}}{\text{horizontal distance}}$ — State the formula for gradient.

$= \dfrac{26}{78}$ — Substitute the values.

$= \dfrac{1}{3}$ — Simplify the fraction where possible.

N4 **Example 7.2**

Gold Street in Shaftsbury and Vale Street in Bristol are both steep streets.

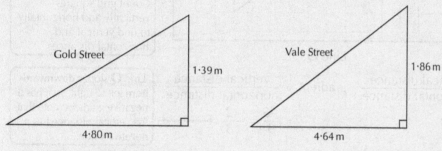

Gold Street

1·39 m

4·80 m

Vale Street

1·86 m

4·64 m

Which street is steeper?

Gold Street:

$\text{gradient} = \dfrac{\text{vertical distance}}{\text{horizontal distance}}$

$= \dfrac{1·39}{4·80}$

$= 0·289...$

$= 0·29 \, (2\,\text{d.p.})$

Vale Street:

$\text{gradient} = \dfrac{\text{vertical distance}}{\text{horizontal distance}}$

$= \dfrac{1·86}{4·64}$

$= 0·400...$

$= 0·40 \, (2\,\text{d.p.})$ — In order to compare gradients, it is easier to express them as decimals.

Vale Street is steeper than Gold Street because it has a greater gradient (0·40 > 0·29). — Give your answer, including your reason (from your calculation).

N4

Example 7.3

Work out the gradients of lines **P** and **Q**.

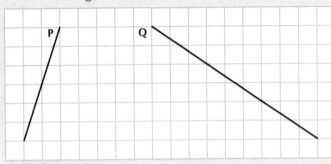

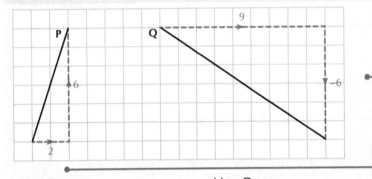

> Choose any two points on the line (it doesn't have to be the start and end points; if you can, use points on the line that cross an intersection of gridlines). Create a right-angled triangle. Count grid squares vertically and horizontally to find vertical and horizontal distances.

Line **P**:

$$\text{gradient} = \frac{\text{vertical distance}}{\text{horizontal distance}}$$

$$= \frac{6}{2} = 3$$

Line **Q**:

$$\text{gradient} = \frac{\text{vertical distance}}{\text{horizontal distance}}$$

$$= \frac{-6}{9} = -\frac{2}{3}$$

> Line **Q** slopes **downwards** from left to right, so it has a **negative** gradient; note that the vertical distance has a negative value.

N4

Example 7.4

Draw a line with a gradient of 2.

A line with a gradient of 2 means that for every 1 unit moved horizontally, 2 units are moved vertically.

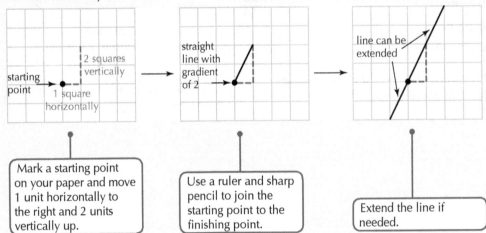

> Mark a starting point on your paper and move 1 unit horizontally to the right and 2 units vertically up.

> Use a ruler and sharp pencil to join the starting point to the finishing point.

> Extend the line if needed.

N4 **Exercise 7A**

1 Calculate the gradient of each slope.

a
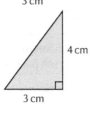
6 cm
3 cm

b
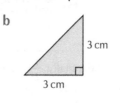
3 cm
3 cm

c
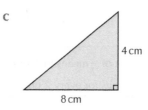
4 cm
8 cm

d
4 cm
3 cm

e
8 cm
10 cm

f
4 cm
12 cm

2 Look at your answers for Question 1.

a Which line is the steepest?

b Which line is the least steep?

3 A car is driving up a hill. It covers 300 horizontal metres and 10 vertical metres.

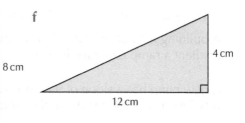

10 m

300 m

Calculate the gradient of the hill.

 4 Calculate the gradients of lines **A–H**.

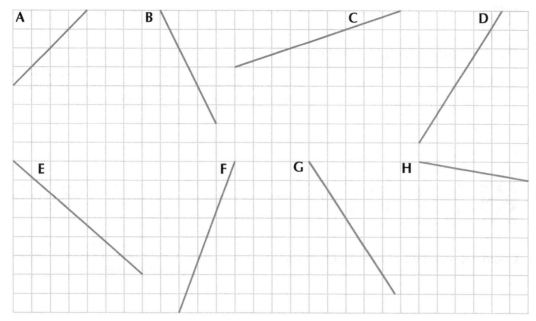

5 Draw lines with the following gradients.

a 1   b 3   c −2

> **Hint** A negative gradient slopes downwards from left to right.

d $\frac{1}{2}$ e $-\frac{3}{4}$ f $-\frac{4}{3}$

> **Hint** A gradient of $\frac{1}{2}$ means that for every 1 unit moved horizontally, $\frac{1}{2}$ unit is moved vertically up. Alternatively, for every 2 units moved horizontally, 1 unit is moved vertically up.

6 For each pair of points, plot the points on a set of coordinate axes, join them to get a straight line and then calculate the gradient of the straight line that you have drawn.

a (1, 3) and (4, 9)   b (2, 4) and (3, 7)   c (−1, 2) and (3, 4)

7 A building regulation states that the maximum gradient a ramp can have is 0.06

> **Hint** Pythagoras' theorem is covered in Chapter 4; see page 43.

A new ramp has a slope of 9 metres and the end of the ramp (at its highest point) is 0·5 metres vertically above the ground.

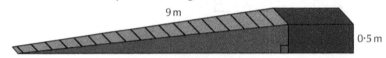

9 m

0·5 m

Does the ramp meet building regulations? Justify your answer.

N4 Drawing a straight-line graph from its equation

You can draw a straight-line graph using its equation. To do this, you need to work out some coordinate points. If you create a table of values you can work out the y-coordinate for each x-coordinate you choose.

N4 Example 7.5

a Complete the table of values for $y = 2x − 1$.

x	0	1	2	3
y				

b Draw the straight line with equation $y = 2x − 1$.

a $x = 0$: $y = 2 \times 0 − 1 = −1$

> To work out the y-value corresponding to each x-value, substitute each of the x-values into the equation $y = 2x − 1$.

 $x = 1$: $y = 2 \times 1 − 1 = 1$

 $x = 2$: $y = 2 \times 2 − 1 = 3$

 $x = 3$: $y = 2 \times 3 − 1 = 5$

x	0	1	2	3
y	−1	1	3	5

> Complete the table of values.

b Coordinates are: (0, −1), (1, 1), (2, 3) and (3, 5)

> Use the table of values to identify the coordinates.

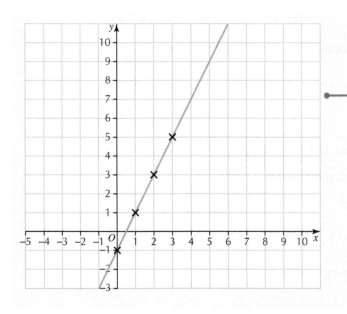

Draw a set of coordinate axes and plot the coordinates. Join the points. The points should be on a straight line. If they are not, then you have made a mistake and will need to check your calculations.

The line can be extended beyond the coordinates that you have plotted.

N4

Example 7.6

Draw the straight line with equation $y = -3x + 1$.

x	0	1	2	3
y	1	−2	−5	−8

Draw and complete a table of values. You can choose any values of x in your table of values but it is often easiest to use $x = 0, 1, 2, 3$.

Coordinates are: (0, 1), (1, −2), (2, −5) and (3, −8)

Use the table of values to identify the coordinates.

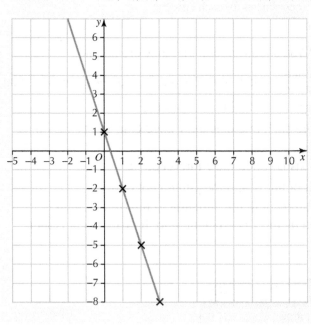

Hint Strictly speaking, you only need to find two coordinates to generate a straight line, but finding three or four gives you an additional way of checking your calculations as shown in Examples 7.5 and 7.6.

N4 **Exercise 7B**

1   For each equation, draw a table of values and then draw the straight line.

   a $y = x$ 　　　　　　**b** $y = 3x$ 　　　　　　**c** $y = -x$

   d $y = \dfrac{1}{2}x$ 　　　　　**e** $y = -5x$

   > **Hint**   In part **d**, use $x = 0, 2, 4, 6$ in the table of values.

★ 2   Draw straight lines with the following equations.

   a $y = x + 2$ 　　　　　**b** $y = 3x + 1$

   c $y = 2x - 3$ 　　　　　**d** $y = \dfrac{1}{2}x + 5$

★ 3   Draw straight lines with the following equations.

   a $y = -x + 3$ 　　　　　**b** $y = -2x - 1$

   c $y = -\dfrac{1}{2}x + 4$ 　　　**d** $y = -4x + 1$

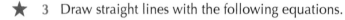

▶N5 The equation of a straight line, $y = mx + c$

The following investigates the equation of a straight line.

On the same set of coordinate axes, draw straight lines with these equations:

- line 1: $y = 2x + 1$
- line 2: $y = 2x + 2$

   > **Hint**   Use a table of values to work out some coordinates for each line.

- line 3: $y = 2x + 3$

What do you notice about the three lines? You should notice that they are parallel.

Use the formula gradient $= \dfrac{\text{vertical distance}}{\text{horizontal distance}}$ to calculate the gradient of the three lines that you have drawn. You should find that the gradient of all three lines is 2.

What are the coordinates of the point where each line cuts the y-axis? Line 1 cuts the y-axis at $(0, 1)$, line 2 at $(0, 2)$, and line 3 at $(0, 3)$.

The point where a straight line cuts the y-axis is called the **y-intercept.**

Look at the equation of line 1, the gradient of line 1 and the y-intercept of line 1.

Can you see values of the gradient and y-intercept in the equation of the straight line? You should notice that the gradient is given by the x-coefficient, and the y-intercept is seen in the term '+ 1'.

Do the same for line 2 and line 3.

> **Important**
>
> The equation of a straight line is written in the form:
>
> $$y = mx + c$$
>
> 　　gradient　　y-intercept
>
> where m is the gradient of the line and c represents the y-intercept.

Example 7.7

For the following straight lines, work out:

   i the gradient   ii the coordinates of the y-intercept.

  a $y = 2x - 3$ **b** $y = -3x + 5$

The equation of a straight line is $y = mx + c$

  a $y = 2x - 3$ $y = mx + c$ ⟵ (Compare the equation with the equation of a straight line.)

    gradient y-intercept

    i Gradient is 2

    ii Coordinates of y-intercept are $(0, -3)$ ⟵ (The x-coordinate of the y-intercept is always zero.)

  b $y = -3x + 5$ $y = mx + c$

    gradient y-intercept

    i Gradient is -3 **ii** Coordinates of y-intercept are $(0, 5)$

Example 7.8

For the following straight lines, work out:

   i the gradient   ii the coordinates of the y-intercept.

  a $x + y = 3$ **b** $4x + 3y - 8 = 0$

  a $x + y = 3$ ⟵ (You need to change the subject in order to get the equation in the form $y = \ldots$)

    $x + y - x = 3 - x$ ⟵ (Subtract x from each side.)

    $y = 3 - x$ or $y = -x + 3$ ⟵ (It is now in the form $y = mx + c$.)

    i Gradient is -1 **ii** Coordinates of y-intercept are $(0, 3)$

  b $4x + 3y - 8 = 0$ ⟵ (You need to make y the subject.)

    $4x + 3y - 8 + 8 = 0 + 8$

      $4x + 3y = 8$

    $4x + 3y - 4x = 8 - 4x$

> **Hint** Look at Chapter 6, pages 88–89, for more information about how to change the subject of a formula.

        $3y = 8 - 4x$

        $\dfrac{3y}{3} = \dfrac{8 - 4x}{3}$

        $y = \dfrac{8}{3} - \dfrac{4}{3}x$ or $y = -\dfrac{4}{3}x + \dfrac{8}{3}$

    i Gradient is $-\dfrac{4}{3}$ **ii** Coordinates of y-intercept are $\left(0, \dfrac{8}{3}\right)$

N5 **Example 7.9**

What is the equation of the straight line shown in the diagram?

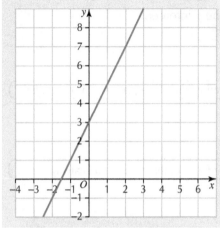

The equation of a straight line is $y = mx + c$ where m is the gradient of the straight line and c is the y-intercept.

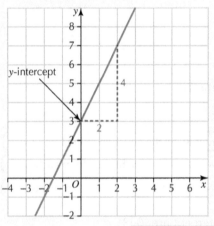

Form a right-angled triangle to help calculate the gradient, m.

$c = 3$ ● From the diagram, you can see that the coordinates of the y-intercept are (0, 3).

gradient, $m = \dfrac{\text{vertical distance}}{\text{horizontal distance}} = \dfrac{4}{2} = 2$ ●

From the diagram, vertical distance = 4 and horizontal distance = 2.

So the equation of the straight line is $y = 2x + 3$.

N5 **Exercise 7C**

▶N5 **1** For each straight line work out:

★

      i   the gradient   ii   the coordinates of the y-intercept.

   a $y = 2x + 1$ **b** $y = -x + 3$ **c** $y = 5x - 1$ **d** $y = \dfrac{1}{2}x + 4$

   e $y = -3x - 1$ **f** $y = 2x - 5$ **g** $y = 1 - \dfrac{4}{3}x$ **h** $y = 2 - \dfrac{1}{3}x$

Hint   Part **g** can be written as $y = -\dfrac{4}{3}x + 1$.

2 For each straight line work out:

    i the gradient   **ii** the coordinates of the *y*-intercept.

    a $x + y = 5$ **b** $2x + y = 7$ **c** $y - 5x = 1$ **d** $2y - 4x = 8$

    e $4x + 3y = 5$ **f** $x + 2y = 8$ **g** $3x + y - 9 = 0$ **h** $3x + 4y - 5 = 0$

3 For each given gradient and *y*-intercept, work out the equation of the straight line.

    a $m = 2$; *y*-intercept = (0, 6)   **b** $m = 5$; *y*-intercept = (0, 3)

    c $m = -1$; *y*-intercept = (0, −8)   **d** $m = \dfrac{2}{3}$; *y*-intercept = (0, 5)

    e $m = 3$; *y*-intercept = $\left(0, \dfrac{1}{2}\right)$ **f** $m = -\dfrac{2}{7}$; *y*-intercept = $\left(0, \dfrac{5}{8}\right)$

> **Hint**   Compare each set of values with the general equation of a straight line, $y = mx + c$.

 4 Work out the equation of each straight line.

a
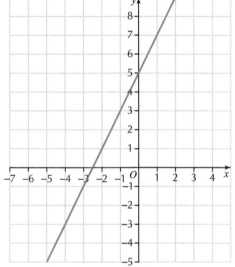

b
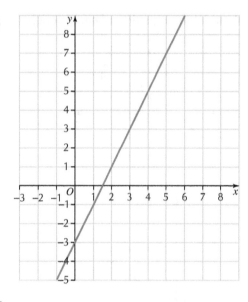

c
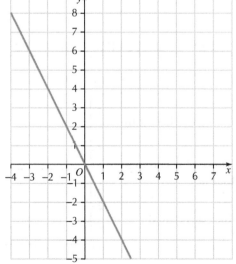

d
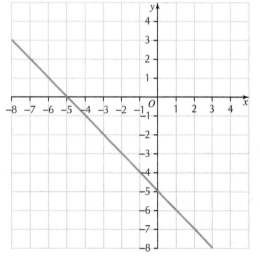

(*continued*)

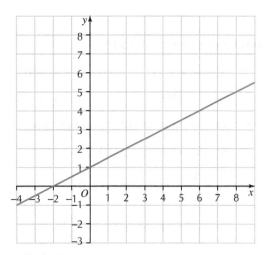

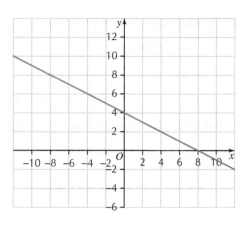

> **Hint** Check the scales on both axes when you are counting horizontal and vertical distances.

5 Work out the equation of each straight line.

a

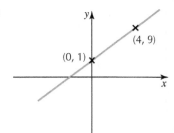

b

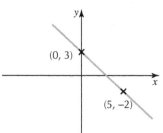

c

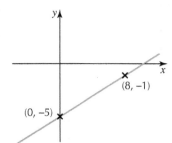

d

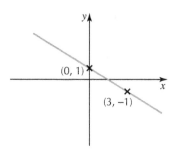

N4 Drawing and interpreting a vertical or a horizontal line

The gradients and equations of horizontal and vertical lines

Horizontal lines
Horizontal lines are flat and run from left to right (like the horizon).

Look at the line shown on the right.

It is not possible to draw a right-angled triangle. There is no change in vertical distance as you move along the line, that is,
change in vertical distance = 0.

However long the line is, any line that has a change of vertical distance of zero will have a gradient of zero.

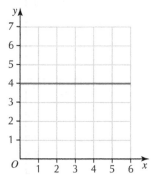

$$\text{gradient} = \frac{\text{vertical distance}}{\text{horizontal distance}} = \frac{0}{\text{anything}} = 0$$

This means that the **gradient of any horizontal line is zero**.
This makes sense because gradient is a measure of steepness, and a horizontal line is flat.

Vertical lines

Vertical lines are straight up and down.

Look at the line shown on the right.

It is not possible to create a right-angled triangle. There is no change in horizontal distance as you move up or down the line, that is, **change in horizontal distance = 0**.

It doesn't matter how long the line is, because any line with a horizontal change of zero will always have the same gradient:

$$\text{gradient} = \frac{\text{vertical distance}}{\text{horizontal distance}} = \frac{\text{anything}}{0} = \text{undefined}$$

The **gradient of any horizontal line is undefined**.

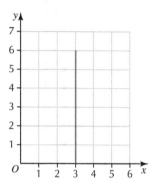

Important

A **vertical** line, drawn on coordinate axes, has equation $x = a$ (where a is the x-coordinate where the line crosses the x-axis).
A **horizontal** line, drawn on coordinate axes, has equation $y = c$ (where c is the y-intercept, that is, where the line crosses the y-axis).

N4

Example 7.10

a Plot these points on a set of coordinate axes, then join them to form a line.

$(-1, 5), (0, 5), (1, 5), (2, 5), (3, 5), (4, 5)$

b Write the equation of the line.

a

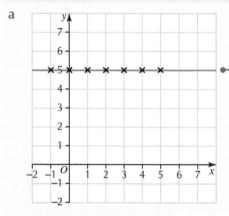

This is a horizontal line, so the gradient is zero. The y-intercept is 5.

b $y = mx + c$ — Write the equation of a straight line.

$y = 0x + 5$ — Substitute the values into the equation.

$y = 5$

The equation of the line is $y = 5$. — For any x-coordinate, the y-coordinate is 5 so the equation of this line is $y = 5$.

N4 **Example 7.11**

   a Plot these points on a set of coordinate axes, then join them to form a line.

   $(-1, -2), (-1, -1), (-1, 0), (-1, 1), (-1, 2), (-1, 3)$

   b Write the equation of the line.

a

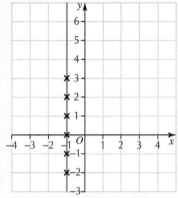

This is a vertical line, so the gradient is *undefined*. The x-intercept is –1.

For any y-coordinate, the x-coordinate is –1 so the equation of this line is $x = -1$.

b The equation of the line is $x = -1$.

N4 **Exercise 7D**

   1 State the equation of each of the following vertical and horizontal lines.

a

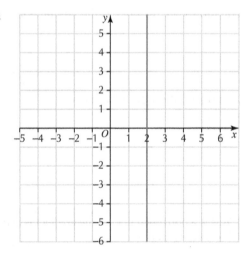

b

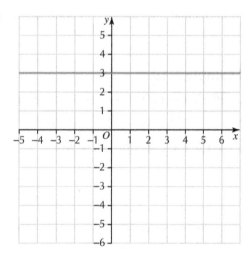

c

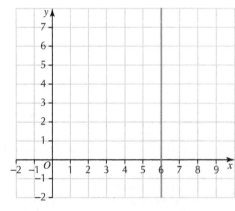

d

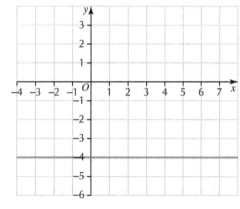

e

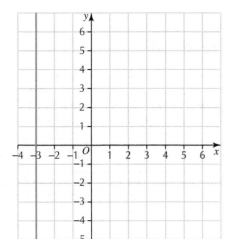

f

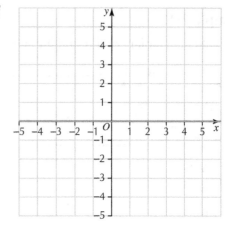

★ 2 On separate coordinate axes, draw straight lines with these equations.

a $x = 3$ b $y = 2$ c $x = -4$

d $y = 6$ e $x = 0\cdot5$ f $y = -1\cdot5$

N5 Calculating the gradient of a straight line, given two points

You can calculate the gradient of a straight line if you are given the coordinates of two points on the line. For example, points A and B have coordinates $(2, 1)$ and $(6, 3)$, respectively.

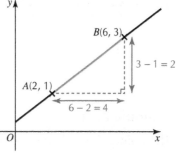

The gradient of the line joining points A and B is:

$$\text{gradient, } m_{AB} = \frac{\text{vertical change}}{\text{horizontal change}}$$

$$= \frac{3 - 1}{6 - 2} = \frac{2}{4} = \frac{1}{2}$$

In general, if you are given two points $A(x_1, y_1)$ and $B(x_2, y_2)$ then you can create a right-angled triangle to help calculate the gradient of the line joining the two points.

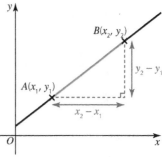

Important

The gradient of the straight line joining points A and B is given by the formula:

$$m_{AB} = \frac{y_2 - y_1}{x_2 - x_1}$$

Example 7.12

Calculate the gradient of the straight line joining the points $A(1, 1)$ and $B(4, 7)$.

$A(1, 1)$ $B(4, 7)$ Identify x_1, y_1, etc.

(x_1, y_1) (x_2, y_2)

Hint   It doesn't matter which point you identify as (x_1, y_1) or (x_2, y_2). The answer will be the same.

$m_{AB} = \dfrac{y_2 - y_1}{x_2 - x_1}$ Write the formula.

$\quad = \dfrac{7 - 1}{4 - 1}$ Substitute the values for x_1, y_1, etc. into the formula.

$\quad = \dfrac{6}{3} = 2$

Example 7.13

Calculate the gradient of the straight line joining $P(-7, 3)$ and $Q(-3, -2)$.

$P(-7, 3)$ and $Q(-3, -2)$ Identify x_1, y_1, etc.

(x_1, y_1) (x_2, y_2)

$m_{PQ} = \dfrac{y_2 - y_1}{x_2 - x_1} = \dfrac{-2 - 3}{-3 - (-7)}$ Be careful when subtracting negative numbers.

$\quad = \dfrac{-5}{4} = -\dfrac{5}{4}$ Write the gradient as $-\dfrac{5}{4}$ and not $\dfrac{-5}{4}$ so that it is clear that the gradient of the line between the two points is negative.

Example 7.14

The point A is $(3, 7)$ and the point B is $(6, k)$. If $m_{AB} = 2$, find k.

$m_{AB} = \dfrac{y_2 - y_1}{x_2 - x_1}$

$2 = \dfrac{k - 7}{6 - 3}$ Substitute the values and create an equation which can be solved to find k.

$2 = \dfrac{k - 7}{3}$

$6 = k - 7$ Multiply both sides by 3.

$k = 13$

Exercise 7E

★ **1** Calculate the gradient of the straight line joining each pair of points.

    a $A(1, 5)$ and $B(2, 9)$ **b** $C(1, 7)$ and $D(2, 10)$

    c $E(0, 1)$ and $F(1, 5)$ **d** $G(-3, -8)$ and $H(3, 4)$

    e $I(0, -4)$ and $J(-1, -9)$ **f** $K(0, -8)$ and $L(-1, -9)$

g $M(-2, 3)$ and $N(-1, 5)$ h $P(1, 7)$ and $Q(2, 7)$

i $R(0, -1)$ and $S(-1, 5)$ j $T(-3, 5)$ and $U(-2, -1)$

2 Look at your answers to Question 1.

     a What comment can you make about the straight line in part **h**?

     b What comment can you make about the straight lines in parts **i** and **j**?

★ 3 Calculate the gradient of the straight line joining each pair of points.

     a $A(1, -7)$ and $B(-5, 0)$

     b $C(4, 1)$ and $D(-6, 4)$

     c $E(6, 9)$ and $F(-4, -2)$

4 Calculate the gradient of the straight line joining each pair of points.

     a $A(-3{\cdot}2, 4)$ and $B(-2, 0{\cdot}4)$

     b $C(1, 3)$ and $D\left(\frac{1}{10}, -\frac{3}{5}\right)$

     c $E\left(\frac{1}{2}, \frac{5}{2}\right)$ and $F\left(\frac{5}{4}, -\frac{3}{2}\right)$

5   a Point A is $(2, 0)$ and point B is $(4, k)$. If $m_{AB} = 3$, find k.

     b Point C is $(4, k)$ and point D is $(8, 11)$. If $m_{CD} = 2$, find k.

     c Point E is $(1, 5)$ and point F is $(3, k)$. If $m_{EF} = -2$, find k.

     d Point G is $(-2, 6)$ and point H is $(8, k)$. If $m_{GH} = \frac{1}{5}$, find k.

N5 Using the formula $y - b = m(x - a)$ to find the equation of a straight line

You can work out the equation of a straight line when you know its gradient, m, and the coordinates of a point on the line (a, b).

> **Important**
>
> The equation of a straight line can be found using the formula:
>
> $$y - b = m(x - a)$$
>
> where m is the gradient and (a, b) is a point on the line.

N5 Example 7.15

Find the equation of the straight line with gradient 2 that passes through the point $(1, 5)$.

$m = 2$

The given point is $(1, 5)$ so $a = 1$ and $b = 5$ •————— (Identify values of m, a and b.)

$y - b = m(x - a)$ •————— (State the formula for the equation of a straight line.)

$y - 5 = 2(x - 1)$ •————— (Substitute values of m, a and b.)

$y - 5 = 2x - 2$ •————— (Multiply out the brackets.)

$\qquad y = 2x + 3$ •————— (Make y the subject, by adding 5 to both sides.)

N5 **Example 7.16**

Find the equation of the straight line that passes through points $A(-2, 0)$ and $B(1, 6)$.

$A(-2, 0)$ and $B(1, 6)$

(x_1, y_1) (x_2, y_2)

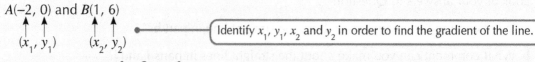

Identify x_1, y_1, x_2 and y_2 in order to find the gradient of the line.

$$m_{AB} = \frac{y_2 - y_1}{x_2 - x_1} = \frac{6 - 0}{1 - (-2)} = \frac{6}{3} = 2$$

Point B is $(1, 6)$ so $a = 1$ and $b = 6$

Choose a point (it doesn't matter which) in order to identify values of a and b.

Hint When deciding which point to choose, it is often easier to pick the point with no fractions or negative numbers.

$y - b = m(x - a)$ — State the formula.

$y - 6 = 2(x - 1)$ — Substitute values of m, a and b.

$y - 6 = 2x - 2$ — Expand the brackets.

$y = 2x + 4$ — Make y the subject.

Check:

$y = 2x + 4$

LHS $= y = 0$

Check your answer by substituting the coordinates of the other point, $A(-2, 0)$, into the equation you have found.

RHS $= 2x + 4 = 2(-2) + 4 = 0$

LHS $=$ RHS ✓

N5 **Example 7.17**

Find the equation of the straight line that passes through points $P(4, 1)$ and $Q(-6, 4)$.

$$m_{AB} = \frac{y_2 - y_1}{x_2 - x_1} = \frac{4 - 1}{-6 - 4}$$

$$= \frac{3}{-10} = -\frac{3}{10}$$

Point P is $(4, 1)$ so $a = 4$ and $b = 1$

$$y - b = m(x - a)$$

$$y - 1 = -\frac{3}{10}(x - 4)$$

$10y - 10 = -3(x - 4)$ — Multiply both sides by 10 to eliminate the fraction.

$10y - 10 = -3x + 12$

$3x + 10y = 22$ or $3x + 10y - 22 = 0$ — Collect like terms.

N5 **Exercise 7F**

★ 1 Find the equation of the straight line with:

   a a gradient of 1 that passes through the point $(5, 5)$

   b a gradient of -1 that passes through the point $(-6, 4)$

   c a gradient of 2 that passes through the point $(9, -3)$

   d a gradient of -3 that passes through the point $(8, -8)$

e a gradient of 4 that passes through the point (−8, 0)

f a gradient of 3 that passes through the point (3, 5)

g a gradient of 3 that passes through the point (1, −6)

h a gradient of 5 that passes through the point (−6, 5)

i a gradient of −2 that passes through the point (7, −3)

j a gradient of 1 and passing through the point (−1, 3)

k a gradient of 10 that passes through the point (−1, −8)

l a gradient of $\frac{1}{3}$ that passes through the point (8, 5).

 2 Find the equation of the straight line that passes through each pair of points.

a (2, 4) and (3, 8) **b** (−1, −5) and (−3, 3) **c** (3, −1) and (7, 3)

d (2, 9) and (7, −1) **e** (−5, 3) and (7, 1) **f** (3, −2) and (18, −5)

g (2, 0) and (2, 8) **h** $\left(\frac{1}{4}, \frac{1}{2}\right)$ and $\left(\frac{1}{2}, \frac{7}{12}\right)$

| Hint | In part **g**, a line with an undefined gradient is vertical. For a reminder about horizontal and vertical lines, see pages 106–109. |

N5 Using and applying function notation $f(x)$

A **function** applies a given rule to one set of numbers to give a new set of numbers.

For example, $f(x) = 3x + 1$ is a function; different x-values can be substituted into it to get a different number out. In this function, the rule is to multiply the number by 3 and add 1, so when $x = 2$, $f(x) = 3 \times 2 + 1 = 7$. This is written as $f(2) = 7$. So when $f(x) = 3x + 1$:

$$f(1) = 3 \times 1 + 1 = 4 \qquad f(5) = 3 \times 5 + 1 = 16 \qquad f(10) = 3 \times 10 + 1 = 31$$

Functions are expressed in terms of their **domain** and their **range**. The domain is the set of input numbers, and the range is the set of output numbers after the function has been applied to the input numbers.

| Hint | A function can be expressed using any letter; it doesn't have to be f. So, $g(x)$, $h(x)$, etc. are also functions. |

N5 Example 7.18

For the function $f(x) = 5x - 3$, find:

a $f(2)$ **b** $f(−4)$

a $f(x) = 5x - 3$

$f(2) = 5 \times 2 - 3$ ———— Substitute 2 for x and then complete the calculation.

$= 10 - 3 = 7$

b $f(x) = 5x - 3$

$f(−4) = 5 \times (−4) - 3$

$= −20 - 3 = −23$

Example 7.19

The function g is given by $g(x) = 2x^2 + 1$.
Find the value of x for which $g(x) = 33$.

$g(x) = 2x^2 + 1$

$33 = 2x^2 + 1$ ●————————————————————— Substitute 33 for $g(x)$.

$2x^2 + 1 = 33$ ●————————————————— Swap LHS and RHS to get the x^2 term on the LHS and then solve for x.

$2x^2 = 32$

$x^2 = 16$

$x = \pm 4$ ●————————————————— Take the square root both sides. Note that both $4 \times 4 = 16$ and $(-4) \times (-4) = 16$.

Exercise 7G

1 For the function $f(x) = x + 4$, find:

   a $f(5)$ b $f(0)$ c $f(-3)$ d $f\left(\dfrac{1}{2}\right)$

2 For the function $f(x) = 3x - 1$, find:

   a $f(4)$ b $f(0)$ c $f(-8)$ d $f(0·5)$

★ 3 For the function $g(x) = 2x - 7$, find:

   a $g(3)$ b $g(-4)$

   c $g(5) - 2$ d $g(-3) + 4$

> **Hint** In part **c**, find $g(5)$ and then subtract 2 from the answer.

4 For the function $g(x) = 7 - 2x$, find:

   a $g(3)$ b $g(-4)$ c $g(-8) + 1$ d $g\left(\dfrac{1}{3}\right)$

★ 5 For the function $h(x) = x^2 + 5$, find:

   a $h(3)$ b $h(0)$ c $h(-2)$ d $h(-10) - 3$

6 The function f is given by $f(x) = 3x + 5$. Find the value of x for which $f(x) = 11$.

★ 7 The function f is given by $f(x) = 4x + 1$. Find the value of x for which $f(x) = -7$.

8 The function g is given by $g(x) = -5x + 3$. Find the value of x for which $g(x) = 13$.

9 The function h is given by $h(x) = 2x^2 + 3$. Find the values of x for which $h(x) = 53$.

10 Functions f and g are given by $f(x) = 3x + 1$ and $g(x) = 2x - 5$.

   Find the value of x for which $f(x) = g(x)$.

11 Functions g and h are given by $g(x) = x^2 + 5x - 2$ and $h(x) = x^2 - 4x + 7$.

   Find the value of x for which $g(x) = h(x)$.

N4

N5

Chapter 7 review

1 Calculate the gradient of each line.

a 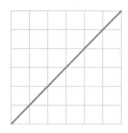   b    c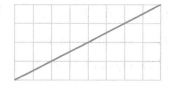

2 State the equations of the following vertical and horizontal lines.

a 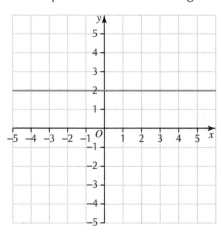   b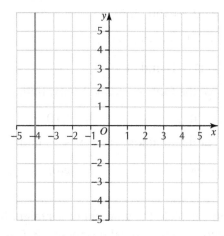

3 On separate coordinate axes, draw a straight line with the given equation.

   a $y = 3x - 1$ b $y = -4x + 2$ c $x = 1$ d $y = -3$

4 For each straight line work out:

   i the gradient   ii the coordinates of the y-intercept.

   a $y = 2x + 5$ b $y = -3x - 4$ c $2x + y = 6$ d $3x - 2y = 5$

5 What is the equation of each straight line?

a 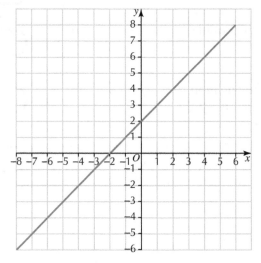   b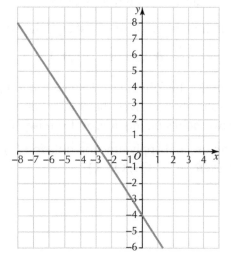

6 Calculate the gradient of the straight line joining each pair of points.

    a $A(1, 3)$ and $B(3, 9)$

    b $C(5, -2)$ and $D(-10, 3)$

    c $E(-1, -4)$ and $F(-8, 6)$

7 Find the equation of the straight line with:

    a a gradient of 3 that passes through the point $(2, -2)$

    b a gradient of -4 that passes through the point $(-3, 1)$

    c a gradient of $\frac{3}{2}$ that passes through the point $(-1, 5)$

8 Find the equation of the straight line that passes through each pair of points.

    a $(2, 4)$ and $(3, 6)$

    b $(-1, -5)$ and $(-3, 3)$

    c $(1, -1)$ and $(3, 6)$

9 For the function $f(x) = 2 - 3x$, find:

    a $f(1)$ b $f(10)$ c $f(-4)$ d $f(5) + 8$

10 For the function, $g(x) = x^2 + 7$, find:

    a $g(5)$ b $g(0)$ c $g(-3)$ d $g(-6) + 5$

- I can calculate the gradient of a straight line from horizontal and vertical distances. ★ Exercise 7A Q4

- I can draw a straight-line graph given its equation. ★ Exercise 7B Q2, Q3 ★ Exercise 7D Q2

- I can identify the gradient and y-intercept from the equation of a straight-line. ★ Exercise 7C Q1, Q2

- I can identify the gradient and y-intercept from a straight-line graph and use this to work out the equation of the line that is drawn. ★ Exercise 7C Q4

- I can find the gradient of a straight line using the formula

    $m = \dfrac{y_2 - y_1}{x_2 - x_1}$ ★ Exercise 7E Q1, Q3

- I can use the formula $y - b = m(x - a)$ to find the equation of a straight line given the gradient and a point on the line or two points on the line. ★ Exercise 7F Q1, Q2

- I can use and apply function notation $f(x)$. ★ Exercise 7G Q3, Q5, Q7